the art
of
ex·plan·ation

the art
of
ex·plan·ation

noun

1. How to Communicate
with Clarity and Confidence

ROS ATKINS

WILDFIRE

First published in 2023 by
WILDFIRE
an imprint of HEADLINE PUBLISHING GROUP

3

Cataloguing in Publication Data is available from the British Library

Hardback ISBN 978 14722 9844 7
Trade paperback ISBN 978 14722 9840 9

Designed and typeset by EM&EN
Printed and bound in Great Britain by Clays Ltd, Elcograf S.p.A.

HEADLINE PUBLISHING GROUP
An Hachette UK Company
Carmelite House
50 Victoria Embankment
London EC4Y 0DZ

www.headline.co.uk
www.hachette.co.uk

To Sara, Alice and Esther

Contents

INTRODUCTION

We all know the feeling. If it's a presentation, maybe people's eyes are drifting, or their phones are coming out. If it's an essay, it's a gnawing sense that the words are there but you're not sure what they add up to. If it's an email, it's no reply – or one that doesn't address the questions that you asked. For whatever reason, the information we want to communicate is neither hitting the mark nor delivering what we were hoping for in return.

Explanation can help with this. Explaining yourself well does not guarantee your desired outcome. You could do it perfectly and still not get the job, make the sale or win the argument. But when you convey what you mean with clarity and impact, you give yourself the best chance of being understood and achieving whatever goals you may have.

Do that in all your interactions – from everyday moments like a quick conversation with your kids' teacher, an email to a colleague or a meeting with a builder, all the way to the biggest moments, such as an interview, an exam, a speech, a pitch, a dissertation, or a difficult conversation – and explaining yourself can be transformational. You are removing the obstacles to the world understanding what you want to say. Get it right and you can radically improve how you communicate.

Explanation is an art. I've spent a lot of the last thirty years thinking about how to do it. Not least because, for most of them, I've been a BBC journalist and presenter. My working life is spent seeing how I can take all the complexities of what's happening in our world and give them shape, clarity and relevance. Not just that, I soon realised that if I wanted to get any new ideas off the ground, I'd need to be able explain them too; each time calibrating how to give them impact with the person or people I needed to persuade of their merits.

That twin desire of wanting to explain the news and to explain my ideas means that most days, for two decades and more, I've pored over every technique I can find to make my explanations better – testing, reviewing, ditching and developing them as I've gone along. Explanation has become a passion. And the more I've explored how to do it better, the more its potential far beyond my journalism has revealed itself.

When done well, explanation helps us establish what we want to say, to find and distil the information that we need to say it – and to calculate the most effective way to give it to people. Explanation is all of this. It encompasses what we say *and* how we say it – and everything that comes within those ambitions.

In recent years, my eyes have opened to how explanation is woven into the fabric of all our experiences. Rather than being a discipline that was making me, I hope, a better journalist, now I see something urgent, powerful and universally applicable. These are just some of the benefits:

- If a teacher can explain the complexities of an equation, the chances of a pupil enjoying and understanding maths shifts too.
- If an email explains persuasively why a decision needs to be taken, there's a greater chance that colleagues will agree.
- If a student can organise information and explain its importance and relevance, their grade will be higher.
- If a candidate can explain why they should get the job, they've a better chance of getting it.
- If a doctor can explain the benefits of a new diet, the chances of a patient sticking to it increase.

- If a government agency can explain how to access its services, people are more likely to use them.

- If a company can explain the merits of its product, it's more likely to make a success of it.

- If an entrepreneur can explain their business idea, investment is more likely.

- If a theatre's website can explain how ticketing works, it'll get fewer emails asking about that.

- If a builder can explain what they're able to do and when, a family is more likely to be satisfied with their work.

The list is endless. Explanation cuts across every aspect of our lives.

From something as fundamental as someone's health, to an important job interview to simply saving time in day-to-day life, quality explanation can make the difference. And, as I've learned of its potential, I've also learned there's no magic bullet to achieve this. Instead, explanation done well is the culmination of a range of complementary actions. Done in sync, they work together to deliver whatever it is you want to say. This book is my effort to share what I've learned about how to do this.

Explaining yourself brilliantly is far from easy. I never take it for granted. Indeed, I'm painfully aware that, with this book, I am setting myself a high bar. I'm sure there will be times when you think I could have been clearer. The reality is that we can always explain ourselves better. This, after all, is an art not a science. But that shouldn't stop us being as attentive to explanation as we can be. Because there can be costs if we don't pay attention.

Every day, I see examples of people who have important things to say and who fail to get them across – politicians who have a new policy but I'm none the wiser after they've talked about it; emails where the nugget of information I need is buried in a mountain of text; job applications where someone is saying they're right for the role but I can't find their evidence to back that up. You'll be able to think of examples too. Being on the receiving end of poor explanation is part of daily life. I'm sure it always has been.

The internet, though, has altered the consequences. We live in an era of infinite information. It is all around us. Writing in the *Boston Globe* in 2020, the academics Todd Rogers and Jessica Lasky-Fink explained, 'The average professional spends 28 per cent of their work week – over eleven hours – reading and answering email. The average person communicates more than ninety times per day by text and more than 100 times per day by email.'[1] When I read this, the sheer volume was both shocking and not shocking at all. These numbers ring true for me. They may well feel familiar to you too.

Todd Rogers and Jessica Lasky-Fink go on: 'Every message received demands attention and time from people who are already too busy.'

It's not just messages that are coming at us. There are many other information sources that we choose to seek out – from streamed TV to podcasts to games to endless websites and apps. The merits or otherwise of our always-on online lives is for another book (in fact, a number have already been written) but for the purpose of explanation, the fact is that the information you want to share is operating in an exceptionally competitive environment. Not only do you need to get people's attention, but then, when you have it, you need to make what you say count. I never lose sight of how difficult that is. My starting point is that the odds are stacked against us.

When I'm making an explainer video, I begin from the position of thinking no one's going to watch it. I don't assume people will be interested in the subject or, if they are, will want my thoughts on it. I also don't assume that if they start watching, they'll keep watching.

To put it another way, my starting point is that a successful explanation – by which I mean identifying the information, shaping it, passing it on and it being understood and, if need be, acted on – is something I am going to have to fight for every step of the way. That doesn't mean jumping up and down if I'm giving a talk or writing an email in large font. Trying to *demand* attention will not get me far at all. Instead, I try to create explanations that are so clear, focused and relevant, they merit attention in and of themselves. Because explanations are a two-way street. Those of us giving them want to communicate and, sometimes, prompt an action too. The recipient also wants this explanation to be good. Bad explanations take longer to consume and don't help us understand what we're being told. They're extremely inefficient. Given all the information coming at us, we don't want to be on the receiving end of poor explanation any more than we want to be doing it ourselves.

All of this, though, remains a work in progress. As the amount of information around us has spiralled, how we communicate has not always risen to the challenge. Once you start spotting poor explanation, you'll see it everywhere. You might be able to think of times you've done it yourself – when, for whatever reason, you didn't manage to give people the information you wanted them to have. Perhaps you left a meeting without clearly making the points you wanted to get across or you produced a briefing paper asking for support on a project but failed to outline what specifically you needed the recipient to do. As you'll be hearing, I can think of many times I've not got it right.

To be clear, these moments aren't necessarily happening because the information we have to share isn't interesting or relevant or important. It may well be all those things. But *having* this high-quality information isn't enough.

Gathering information is just the start. From there we need to consider where that information is going, how much space there is for it, which parts of it matter most – and how we're going to make it as consumable as possible.

If I'm reporting or presenting a news story for the BBC, I may have all or some of the information on that subject. But what value does that information have if I can't give it to people in a way that makes them want to watch and which makes sense to them? Indeed, if I passed on every piece of reliable information available to me, I may well undermine my chances of explaining the *most important* aspects of it. For me, one of the revelations of recent years was when I started to understand just how many things I was doing that actively got in the way of me explaining myself.

This doesn't just apply to the news. It applies to how we explain ourselves in many parts of our lives. The choices we make about which information to pass on and how we pass it on can, if we're not careful, work against what we're trying to achieve.

Or let's look at this in a more positive way: imagine the possibilities when you can identify *the best and most relevant* information and give it to the people who you'd like it to reach as effectively as possible. I've seen what happens if you can get it right.

Explanation is a potent mix of clarity of purpose, clarity of language and essential information – all calibrated for an intended audience. Those who do it well stand out.

On a practical level, it can help to get things done and it can save you and others a lot of time. On a more fundamental level, effective explanation can alter how you interact with people and how you present yourself. A good explanation can be the difference, which is why it's a preoccupation and a passion of mine.

Over the past thirty years, I've developed a system of explanation that helps me identify information, organise it and then communicate it in, I hope, a clear, concise and comprehensible way. In this book, I'm going to share that system with you. It can be adapted to circumstance. It can be used for everything from extensive preparation for a formal presentation or an exam to the thirty seconds before an unexpected conversation with a colleague. It's a system that I've built my career on and which I still use every day. I'd be lost without it, to be honest. And just as it's helped me, I hope it will help you too.

DISCOVERING THE POWER OF EXPLANATION

My system of explanation wasn't something I worked up out of curiosity. It was born out of necessity. It was my effort to find a way out of several situations that felt, for a time, overwhelming.

Back in the 1990s, two years after I'd graduated, I was unemployed, living in Cambridge and doing a far better job of watching *The Jerry Springer Show* than I was of finding work. I was low on confidence, low on cash and low on ideas for what to do about it.

I did, though, have one trump card to play. After college, I'd gone to live in South Africa with my then girlfriend.

This was a time when Nelson Mandela was President, when the post-apartheid possibilities were still in the first flush of being explored and when crime was on everyone's minds. I landed a gig as a policy researcher and, each day during my drive from the suburbs to our offices downtown, I'd have the coins ready to buy one of the daily newspapers offered by the hawkers at the traffic lights. As soon as I was in, I'd lay the newspaper out on my desk and read it front to back. Carjackings; police reform;

South Africa making the 1998 football World Cup; the British Lions rugby team paying a visit; the rise of Kwaito (think house music but slower); Mandela at the helm. I'd always wanted to be a journalist and South Africa had so many stories to tell.

At the weekend, I got into the routine of wandering down the road and buying the *Sunday Independent*. At the time, it was owned by the same company that owned the *Independent* titles in the UK. As someone who, as a teenager, had feasted on columns by William Leith and Zoë Heller and who'd cut out photos from the *Independent on Sunday* for my bedroom wall, I was instantly drawn to its South African iteration.

The editor of the *Sunday Independent* was John Battersby. Sitting behind a big desk, he had the demeanour of a man who'd seen a lot of stories and was still very interested in the next one. To my surprise, he'd run an opinion piece of mine about policing and, a few months later, I called to ask if I could come and help with the paper at weekends. John generously agreed and, for a few months, as my relationship and my visa both ran down to zero, on Friday and Saturday nights, I'd get involved with whatever needed to be done. I also DJed quite a lot then and so, for a short while, I would be a journalist until 10 p.m. before driving to the club where I played each week. It's a combination I've never quite managed to match since.

This set-up lasted for a while and, when my time came to move back to the UK, John kindly wrote me a reference to take with me. This was my trump card. The letter was addressed to his colleagues at the *Independent* in London.

In the summer of 1998, I sent it off. A few days later, I got a call asking me to come in. This was a gold-plated chance at a time when I was failing to generate any others.

In the days running up to the meeting, I read any number of articles while listening to hours of Radio 5 Live. I'd done my homework on the news. I was far less certain of what to wear.

And so, decked out in a smart casual outfit that accurately told the story of a young man unsure of what was required of him, I boarded the train south.

At the time, the *Independent* was based in Canary Wharf in London. I took the lift up and up and was ushered into a corner office with a view matching the seniority of the people there.

'So what do you want to do for us?' the man sitting behind the desk asked. That I can't tell you precisely who this person was is a measure of how much I've come to want to forget those few minutes.

I know I spoke because I can recall doing so. But what I offered was a disastrous melange of poorly defined hopes and observations.

I had no idea what I needed to say, why I was saying it, how I was going to say it, who I was saying it to or what I wanted from them in response. My strongest memory of the day is the feeling of my chance slipping away. Not at the end of the conversation, but right then in the middle of the first answer. I can feel it now as I write. As I talked, I realised I wasn't explaining myself and that it would cost me. It did. The energy went out of the meeting within minutes and, before long, I was back in the lift. They were kind enough to not write me off there and then. But I never got to work for a paper that I admired.

This stung badly. Instead of working in Canary Wharf, I took a part-time job in a shop selling coffee and coffee machines. For several months, my life was a mix of measuring bags of Colombian beans, daytime TV, some DJing and the occasional game of squash. There was no shortage of time to reflect on why I was struggling to get going.

The flat in Cambridge that I shared with my mate Ollie was only a few hundred metres from the college where I'd studied history. And as I considered my failure to make the opportunity at the *Independent* count, it was to my experience as a student that I turned for a better way to prepare for those moments and, more broadly, for how to plot a way out of the rut that I was in.

The method of teaching history at Cambridge in the nineties was as uncomplicated as it was bracing. Each week, you'd be given an essay question on a subject you either knew nothing or very little about. With the question would come a reading list of anywhere between ten and thirty books and a 'see you next week.' There were lectures but, in an act of academic purity, they were often on subjects that had little to do with the essays we were being set. It really was just about you, the question and the reading list – with the added incentive that, in a week's time, you'd need to hand over a few thousand words and then spend an hour discussing the subject with an academic who'd made it their life's work.

This seemed to me to be entirely unrealistic without a plan. I don't mean an essay plan, though I would need one of those too. No, this required a plan for collecting, processing and then reshaping this new and plentiful information into a form I could confidently use both to explain the subject and to form arguments around it. Maybe others could just read the books, take some notes and identify the issues. I, though, needed more than that and, in those first intimidating weeks of my first year at university, I started to experiment with what might work.

The system I came up with is still largely the one I use today. What I didn't see at the time was that something that helped me explain and analyse nineteenth-century socialism or gender, body and food in late medieval European mysticism (it was nothing if not a varied course) could also help me in many areas of my life – not least in selling myself when I needed to most.

By 2001, I'd escaped the clutches of Jerry Springer and landed a couple of decent jobs editing websites. But the dot-com bubble would soon burst, and I was made redundant. That summer of unemployment, I scoured each Monday's *Media Guardian* for the latest job ads as I became increasingly desperate for something to come up.

I can remember seeing the listing for a producer on Radio 5 Live like it was yesterday. I had a pit in my stomach as I thought about how much I wanted this job. I was an avid 5 Live listener and, if I could have picked anywhere to be a journalist, it would have been there.

I can also remember the room of the interview just as clearly as I can recall how it felt to be in there. I had learned my lesson from my trip to the *Independent*. By this time, I'd realised that the system that had helped me navigate university could help me navigate the most important interview of my life. Afterwards, I was as sure as I could be that I'd got it. I knew I'd managed to clearly and precisely explain what I wanted to say. I'd also successfully managed to both answer the questions I'd been asked and, in doing so, offload a lot of pre-prepared information.

Explanation helped me get into the BBC and, over twenty years on, it's now the foundation of the journalism I do there and of the opportunities I've been able to create. The system I came up with at university helped me navigate my degree, it helped me get a job that I loved, and then helped me to do that job.

It helps me in many ways outside of the newsroom too.

I recently ended up in hospital for five days. I'll spare you the whole story, but I saw various heart specialists, and, for a little while, it was worrying. Because of COVID-19, there were no visitors and so, in between having a raft of tests and doing a decent amount of fretting, I was having to manage all the meetings with the doctors. They'd often happen at short notice and, when I was told one was imminent, I'd grab my pen and pad, pull over the table to the bed and quickly work through the same explanation process I use at work. It helped me organise what I wanted to say, what I wanted to ask and what I wanted to ask for. It was particularly helpful to do this when I was far from my best. I've used the same approach repeatedly when going back to the hospital for follow-up appointments. That's one example; I could give you so many more.

I used the same system of explanation to prepare to talk to my literary agent Will Francis when I wanted to see if he'd take me on. I used it before we talked to Wildfire, who have published this book. I'm using it right now as I write this. I use it all the time.

My work on this system began as an inexperienced teenager from Cornwall sitting in Cambridge wondering how on earth to tackle my first essay. It's evolved into something multipurpose; something that I use every day. My hope is that it'll be of use to you too.

THE CENTRALITY OF EXPLANATION

In 2010, New York University and non-profit news organisation ProPublica teamed up for a 'joint project to experiment with new ways of doing "news explainers"'. I was unaware of this at the time, but, looking back, the aspirations laid out connect directly to the reasons I am writing this book – and to why explanation is central to so many aspects of our lives.

The press release tells us: 'Bringing clarity to complex systems so that non-specialists can understand them is the "art" of the explainer.' New York University Professor Jay Rosen was leading the project and is quoted: 'Good explainers are engaging, not only informative.' He goes on to outline how an explainer 'addresses a gap in your understanding: the lack of essential background knowledge'.[2]

I like all of this: the emphasis on being informative *and* engaging, on making complex subjects accessible and on 'essential background knowledge' – it is this context that helps us understand *why* something matters.

I'd take all those qualities. But I would also offer a broader definition to go with them. For me, a good explanation contains all the information the person or people I'm addressing need to know on the given subject.

Sometimes, that means no more than several simple details. As I'm writing this, we're organising our younger daughter Esther's birthday party. The other parents will need to know when it starts, how long it lasts, where we're going and so on. That information has to be clear and easy to find, and then our work is done (at least until the party starts . . .).

If, however, I was helping to organise a week-long school trip, much more detail would be required.

Or when I'm at work producing a news story, what the audience needs to know may include the latest development and some relevant historical background. We may also need to include political reaction or new data.

Whether it's a news story, a kids' activity or everything in between, ask yourself, 'What do the people I'm speaking to *need* to know?'

This question is very close to my heart. Around 2010 – the same time ProPublica and Jay Rosen were doing their work on explanation – I began to shift from radio presenting to TV and started to present a lot of classic half-hour TV news bulletins. You'll be familiar with the format – headlines, pre-made reports, live 'two-ways' with correspondents, the most interesting pictures of the day, occasionally an interview and so on. It's a format that's found huge success for decades and remains the dominant way that TV news is presented. And as I sat in the studio learning this new medium, it was impossible not to notice the difference between the news I could get through my phone and the news I was giving to the viewers.

On my phone, I could move easily between a vast array of sources and types of content. Let's imagine a bomb had gone off in a city: I could read a tweet from an eyewitness, watch some local TV coverage on YouTube, check a local map, read an article on existing tensions in this city, look at a social media feed of stills from a photographer, check live video feeds from news agencies, follow a reporter from a rival network positioned by the police cordon and read reaction from the local authorities. That list isn't even exhaustive. I could also move between those sources of information as I chose. This type of real-time multi-source collation and consumption – a news-focused form of browsing, if you like – was rapidly becoming central to how many of us followed the news via our phones and computers.

Sitting in the TV studio, it would be possible *in time* to include all those sources in our reports. In that moment, though, it would be very hard. TV is fiendishly complicated to produce, and our production systems meant getting this array of information to the viewer rapidly was all but impossible. These practical considerations meant there was a limit to what we could include *immediately*. As such there was a risk that a viewer faced with the choice of watching me or looking at their phone, might do the latter and switch me off.

This seemed both a problem and an opportunity. The problem is clear: if someone can get better information from their phone, they might not turn to me or my colleagues. The opportunity was that, increasingly, in the words of one former editor, 'We are making news for people who already know the news.' That a bomb has gone off somewhere is a fact that rapidly becomes pervasive – from news sites to social media to WhatsApp and Messenger groups. People were not turning on the television to be told that something had happened – they were turning it on for detail, context, live pictures, curation and analysis. 'Perhaps we could change how we offer that,' I pondered as I learned the TV ropes.

In the coming months, I started to work up an idea for a TV bulletin that could include any piece of information we felt had value, regardless of its source or form. It wouldn't mean we'd use every piece of useful information – that would make for a very long and ineffective show – but now we'd have the option. And we'd also be able to do it at speed.

Most of you will have used livepages at some time or another. News sites run them on big or breaking stories, others for football matches and so on. They often have one person at the helm who posts updates and information from a range of sources. They include whatever they think will help you understand what's happening. Put simply, my idea back in the early 2010s was for a broadcast version of a livepage with me, the presenter, playing the role of the livepage host. I loved the idea of constructing a story or an issue in front of the viewer and saying to them: 'Let's start with this – but to understand that, you also need to know this and this – and to understand those things, bear *this* in mind too.' I would build the story in front of them and take them through it step by step until by the end they had the full picture. That was the theory anyway.

The BBC went for it. We called it *Outside Source* (an 'outside source' is the technical name we use for a video or audio feed coming into the newsroom; it was also a nod to the fact we'd use information from any source, BBC or otherwise). The brilliant BBC designers, directors and engineers then built us a touchscreen that allowed us to pull in any type of content on a story. We were now in a position where we could include anything we liked. This, I felt, gave us a much better chance of delivering high-impact explanation in a live TV environment.

When we launched, the touchscreen raised a few eyebrows and, with a certain predictability, the tweets suggesting it was a 'gimmick' duly arrived. I got it – TV has a long history of using technology because it can, rather than for any great purpose. The tweets, though, petered out quickly because, whether you liked the show or not, it was clear the screen wasn't a gimmick. It was letting me explain stories in a different way.

The screen's now gone – replaced by a new suite of graphics that offers us the same options but looks better on a mobile phone. But the idea behind it remains.

Explanation is about distilling and sharing, effectively, all the essential information on a given subject. That idea still underpins all the TV news I present and how I communicate in many areas well beyond the news.

If I take a step back, I can also see how explanation wasn't only central to the idea of *Outside Source*, it was also central to it happening in the first place. Before I could worry about whether it'd be a successful format, I needed to make the case to my bosses at the BBC that this type of explanation was worth investing in. I was confident it was a good idea, but having a good idea alone isn't enough. Just as with a news story, this is about both *what* you have to say and *how* you say it.

It's fair to say that my first pitch was not my best, nor indeed were others that followed. But over several years, I thought about what was working and, in particular, what wasn't. I honed how I made the case. I thought carefully about how I described the idea, how I addressed questions and concerns and why I thought it had a chance of success. Not just that, I also thought about where and how I made that case – sometimes this was best done in written form, sometimes in an informal conversation, sometimes with a more formal presentation. This didn't guarantee that the BBC would say yes, but by 2013, it did.

My experience of getting *Outside Source* commissioned taught me a lot about the power of explanation in promoting an idea or an ambition. In 2016, I started something that would show me this on a far greater scale.

EXAMPLE

The 50:50 Project is an initiative that works to increase the diversity of contributors in media content. It began life on *Outside Source* before spreading across BBC News, then across almost all BBC content before expanding still further into organisations in over thirty countries.

Initially 50:50 was focused on increasing the number of female contributors in the BBC's journalism. Back then, I felt that we were stuck in a constant state of trying: representing women equally in our journalism had long been accepted as a desirable and important goal but we'd perhaps also accepted that it wasn't possible. The very real obstacles to achieving equal representation had morphed into reasons for not being able to get there. I wanted to see if we could think of equal representation as non-negotiable, something we expected of ourselves. Just as we did with impartiality, high production values or hitting deadlines. My hope was that if we could prove that achieving a level of 50 per cent female contributors was possible on a news programme like *Outside Source*, perhaps that could help encourage others to take part too.

I very much began with the assumption that I wouldn't succeed with this. Not because I didn't think it was possible to reach 50 per cent female contributors. Nor because I didn't think people cared (needless to say, many others before me had done brilliant work in this area). But because I wasn't confident that I'd be able to persuade very busy teams to take on a new way of approaching this. And this was *all* about persuasion. I'm a presenter. And while we're certainly not shrinking violets, presenters are not managers or editors and so can't tell anyone what to do (nor should we be able to!). Just as with pitching *Outside Source*, this would come down to how well I could make the case.

50:50 uses a simple system of self-monitoring. With the help of the show's lead producers, Jonathan Yerushalmy and Rebecca Bailey, we successfully trialled it on *Outside Source*; I now needed to see if I could get others involved.

First and foremost, I had to explain to them what that meant in practice. I worked and worked at making both the system and my explanation of it as simple and as credible as possible. I also made a list of every question that I thought I'd get asked and wrote down my answers too.

They included:

- *Do we have to do it? (No)*
- *Can we stop if we don't like it? (Yes)*
- *Is it a quota? (No)*
- *How long does it take? (Under 5 minutes a day)*

Other questions required more detailed answers:

- *Does the audience care?*
- *What do we have to do each day?*
- *Why are you doing this?*

Several times I sought advice to make sure my answers added up and, if they didn't, I altered them. Only when I was happy with my answers did I start talking to colleagues. And when I did, those conversations went well and, slowly but surely, 50:50 started to grow.

Talk by talk, meeting by meeting, I got myself to the point where I don't think I could have been clearer at explaining what I was doing and why I was doing it. I also learned to adjust my words depending on who I was talking to – understandably, different teams and different managers would have different questions and concerns. None of this preparation meant people would say yes (and some didn't), but, once more, I was giving an idea its best chance.

This is the power of explanation. Without the time spent thinking on how to explain my ideas, I don't think either *Outside Source* or 50:50 would have got off the ground. I don't think my explainer videos would have happened either. We only started making them after I spent several months in 2019 thinking through not just what they could be but how I could make the case for them too. Indeed, these two challenges are intertwined. The need to persuade others of the idea helped me to improve the idea.

Explanation is relevant whatever you're making the case for. It not only improves how you promote an idea, a request or a point of view – it will also improve the very thing you're promoting as well. In turn, if what you're promoting and how you're promoting it both improve, the chances of a desirable outcome increase.

That's good news if, like me, you're an individual trying to make the case to an organisation.

It's also good news if you're a leader making a case to staff. As the academic and business strategist Professor Lucy Kueng puts it, 'Less clarity means less buy-in.'[3] Or to turn that round, explain yourself with clarity and colleagues are much more likely to be persuaded.

Quality explanation can help with all this and much more. It can be the difference between an idea becoming something real and, well, staying an idea; between a suggestion being persuasive or not; between someone finding what you're saying helpful or not; between someone being well informed by you or not. It's central to how we communicate.

THE ANSWERS ARE ALL AROUND US

There are many ways to explain something and some will work much better for you than others. The good news is that, while we are surrounded by examples of poor explanation, we will also encounter many people who are getting it right. As we wrestle with how we can all do this better, it turns out some of the best lessons we can learn about communication are all around us. I've been avidly collecting them for years.

In 2003, I joined the BBC World Service. Back then, it was still in its home of many years, Bush House in London. Situated at the top of the curve of Aldwych, flanked by a broad tree-lined pavement and dominating its surrounds, the building was very much part of the World Service's identity. As if to show its intention to broadcast far and wide, its towering entrance stares all the way up the avenue opposite. When you arrive, you walk between two enormous columns; inscribed above them are the words 'To the friendship of English-speaking peoples'. Set yourself up like this and you need to back it up. And the World Service did and does.

I joined as a producer. But one day, in late 2004, my editor walked over, said a presenter had missed their train and asked if I would fill in.

'2300 hours GMT. BBC World Service. This is *The World Today.*'

Those were my first words. Grand in their own way, and terrifying too. From there the hard work started – explaining George W. Bush's second term as President, the death of Yasser

Arafat, the end of Tony Blair's time as Prime Minister, the war in Iraq, the 2005 earthquake in Pakistan. There were hundreds of stories to be reported to people who may have known a lot about them but equally may not have known much and were eager to know more. And all the while, you knew you needed to keep people listening too. You may be able to explain the finer points of Iran–US tensions under presidents Ahmadinejad and Bush, but can you do it in a way that is both accurate and takes people with you?

I started to study the minutiae of communicating and broadcasting. It turned out a lot of what I needed to know was right in front of me. I could see and hear who was explaining and communicating well. Listening to the radio, sometimes, I could tell what the presenter had done; other times I could tell it worked but wasn't sure why. In those cases, if they were at the BBC, I could ask. I began capturing everything I could. If a turn of phrase, a sentence structure or a type of delivery had worked, I'd note it down and try to use it myself.

This complemented advice I'd been given by Julian Worricker, a BBC presenter I'd listened to religiously, well before I'd ever imagined I'd become one myself. Feeling all at sea before presenting my first show on 5 Live, Julian had kindly responded to an email asking for help. 'Keep listening back,' he told me. I still do. By chance or intent, we all do things that communicate well and things that definitely don't. In the case of radio and TV, we have the luxury of actually being able to go and listen or watch back. In regular life, we can't do that, but we can still take time to consider how something has gone. If we don't consciously think about what's working and what isn't, it's far harder to do it better next time.

Those early days as a presenter saw me constantly bothering colleagues for advice (and I remain grateful to them). I became and remain a magpie - looking far and wide for techniques and perspectives that can help me to be better at explanation. Not only from news, I hasten to add. Music, entertainment, comedy, business and academia all offered inspiration. Whenever I saw people giving their communication clarity and impact, I'd look at what they were doing and make a note. If it worked for me, I'd add it into the system that I'd started at university. As I would with things that I thought I'd done well myself.

My passion for this was initially rooted in a simple fact for news broadcasters. If you want to give your work the best chance of being consumed and of being helpful, you need to be clear on what you're saying – and you need people to be listening or watching. If they turn off, it doesn't matter what you're saying.

Now, of course I understand that you may well not be on the TV or radio, but the same considerations are in play for all of us in everyday life. How do you keep people tuned in to what you're saying? And while they are tuned in, how do you make sure what you say is impactful and useful?

This is the nub of it. It's the reason this matters so much to me. If someone's not listening to you or if they're not taking in what you'd like them to hear, it's reasonable to ask why are we spending time and energy telling them? What is the purpose of communicating if we're not giving ourselves the best chance of *being heard and understood*?

What started as something driven by a desire to explain the news has morphed into something more profound: a mission to understand how to effectively explain. It's a mission that applies to subjects I know intimately and subjects I knew nothing of twenty-four hours ago. I should warn you, sometimes the former can be harder than the latter.

My magpie approach to improving how I explain myself has meant a whole range of people, moments, issues and events have become attached in my mind to the techniques and processes that I use. As you'll see in this book, I'm always on the lookout for inspiration and will take it wherever I find it. My hope is that reading this book will inspire you to do the same.

HOW TO USE THIS BOOK

Until I was around forty, I was a useless cook. I was interested in food but, beyond a basic chilli con carne or curry, I had little to offer. I was intimidated by the whole process of cooking.

Then, my wife, Sara, gave me a book called *Persiana* by Sabrina Ghayour. The recipes were short, the ingredients lists relatively so and the language straightforward. Gingerly at first, I began trying some of them out. To make it feel less daunting, I'd concentrate on each step rather than worry about the end product. To my delight, lo and behold, I seemed to be doing it and the food was, dare I say it, more often than not delicious. (I also went from being unaware of the existence of pomegranate molasses to getting through an awful lot of it.)

With cooking, you might have some ingredients and then decide what to make. Or you might get a cookery book out, choose a dish, get the ingredients and let the recipe take you through it.

With explanation there are similar questions – what information do you have? What information do you need? How many people is it for?

This book is designed to help you whatever stage of the process you've reached.

If you are confident at explanation, hopefully I can offer some structure, inspiration and techniques to add to what you already do.

If, though, like me with cooking, sometimes you're just not sure where and how to start with explaining yourself – this book is designed to give you processes to help you through that. And if you follow the system step by step, I'm hopeful you will soon gain confidence and get the results you want.

I've divided the book into five sections.

1. **The Anatomy of a Good Explanation** First things first, we need to know what we're aiming for. Here you'll find my ten criteria of a good explanation. These are my guide, whatever explanation I'm working on.

2. **Know Your Audience** No explanation exists in a vacuum. We can't get it right unless we factor in who we're talking to. Here we'll look at how to assess your audience and how to tailor what you say according to who they are.

3. **Seven-Step Explanation** This is the full version of my approach to explanation. You'll find detail on what to do and why I'm suggesting you do it. This can be used for any sizeable undertakings – a presentation, speech, dissertation, strategy document or sales pitch – as well as commitments such as a written briefing, an essay or a talk.

4. **Seven-Step Dynamic Explanation** It's just as important that we can explain ourselves in fluid and unpredictable situations. In other words, talking to other people!

   This section builds on everything we'll already have gone through and looks at how we can bring the clarity that we have in a controlled scenario to a dynamic one.

   If you're preparing for an interview, a negotiation, an important meeting, a conference panel, a staff Q&A, a discussion, or any situation where it's vital that you explain yourself but you can't be sure what you'll be asked, this section is for you. You can be more in control than you might imagine.

5. **Quick Explanations: Verbal and Written** Both versions of the Seven-Step Explanation are designed to guide

you through complex and multifaceted explanations. However, there are many other circumstances when we need to explain ourselves without much notice and in a much shorter form. There are moments every day where we need to explain ourselves quickly. In this section, I've looked at how to approach quick written or verbal explanations, such as an email, text messages, a short-notice work meeting or a five-minute conversation with a plumber, a teacher or a client. While more ambitious explanations can really impact important moments in your life, in their own way these shorter day-to-day explanations can be equally transformative. There are so many of them that if we get them right, it can make a huge difference.

WITH ALL THAT SAID . . .

The purpose of the book is not to try to help you to anchor an election programme or report from a summit (although there's no reason why you shouldn't!). The purpose is to provide a range of ideas, tools and techniques that you can use in your life.

I've tried to outline the kind of situations that each approach may work for but, just as with cookbooks, everyone sometimes goes off-piste and does their own thing and I hope you will too. Our goal is to work out what we want to say, who we want to say it to and then get that information across as effectively as possible. There's no fixed way of reaching that destination. But I've found it immeasurably helpful to have a system that underpins and guides my efforts to do this. Fingers crossed, you will too.

Having set all that out, let's get into it – beginning with the fundamentals of what we're working towards.

1

THE ANATOMY OF A GOOD EXPLANATION

There are ten attributes that I am looking for in an explanation.

1. SIMPLICITY
2. ESSENTIAL DETAIL
3. COMPLEXITY
4. EFFICIENCY
5. PRECISION
6. CONTEXT
7. NO DISTRACTIONS
8. ENGAGING
9. USEFUL
10. CLARITY OF PURPOSE

I have these ten points in mind when I need to communicate. They guide me towards explaining myself better and over time I've learned to instinctively check for them. The system we'll work through later deals with the practicalities of explanation. These attributes are the principles that underpin all of that. Most of the problems any of us have with getting our message across are connected to not managing to do these.

In its simplest form, this is what I'm trying to do when I'm explaining myself:

- I want to provide all the information that someone needs from me or that I would like to give them.

- And I want to give them that information in a way that gives it the best chance of being consumed and understood.

Let's look at the ten elements that will help me do that.

1. SIMPLICITY

Allan Little is a giant of BBC journalism and one of the great script writers of his generation. In a brilliant training video I found in the BBC archive he tells us:

'Simplicity is the key to understanding. Short words in short sentences present the listener or reader with the fewest obstacles to comprehension.'

When I first heard this sentence, it felt like a switch being flicked. Not because I was unaware of the importance of simple language and short sentences. I'd been focused on them for some time. No, the switch was flicked by Allan's phrase: 'obstacles to comprehension.' There I was, quietly watching a training course, and now I felt like jumping up from my desk and telling whoever would listen what I'd just heard. I didn't, I hasten to add, but I hadn't overestimated the importance of the moment to me.

Before, when I thought about simplicity, I saw overcomplication as a stylistic issue – something I wasn't keen on but could live with. When preparing a news script, I'd concern myself with whether the main facts and necessary context were there. I wouldn't be worrying too much if I'd also included, in passing, the name of a mid-ranking minister who wasn't central to the story or a series of statistics that was reasonably interesting but not essential. So long as the main facts were there, 'no harm done' was my calculation. How wrong I was.

I was now seeing these unnecessary details as a direct threat to what I was actually trying to do. I adapted Allan's phrase and started referring to 'obstacles to understanding.' This concept now shapes many of my calculations when I am working on an explanation.

The idea here is that I am holding all this information that I've carefully assembled – and, at the same time, within that information there may be words, facts and phrases that are obstacles to people understanding what I'm trying to say. Think of it in these terms, and these words, facts and phrases that you have stop being benign or neutral – they can do harm to your explanation.

Suddenly, everywhere I looked, there were obstacles to understanding. I started to realise how much of what I was saying added *nothing* to what I was trying to communicate. Worse still, it was actively reducing the chances of me successfully explaining myself. From that moment on, my commitment to simplicity moved up several gears.

These days, in my sights are superfluous adjectives, obscure or complex vocabulary, unnecessary detail, long sentences. They all have to go.

The New Zealand government appears to share these ambitions. In 2022, it proposed a Plain Language Bill (it was passed in October that year). I was sent the story by now former BBC colleague and *Guardian* journalist Jonathan Yerushalmy.

The *Guardian*'s report said that the law 'will require government communications to the public be "clear, concise, well organised, and audience-appropriate"'.[4]

Words after my own heart.

The article quotes Lynda Harris, who created a plain language award in New Zealand: 'Government communications decide the most intimate and important parts of people's lives: their immigration status, divorce papers, entitlements to welfare payments or ability to build a home.'

We also hear from the MP Rachel Boyack, who argues that, 'When governments communicate in ways that people don't understand, it can lead to people not engaging with services that are available to them, losing trust in government and not being able to participate fully in society.'

Whether a law is needed to monitor how government agencies communicate is for others to decide. But what is beyond doubt is that in these circumstances, clarity of explanation isn't an added luxury: this can directly affect people's lives.

In the 2020 article by Professor Todd Rogers and Jessica Lasky-Fink that I quoted earlier, they are unequivocal on this: 'Simplifying language can increase the likelihood that readers will pay attention to, understand and follow through on your message.' They also advocate 'commonly spoken language.' In other words, communicate as a person would. As *you* would.

There is, though, a danger lurking here. In our pursuit of simplicity and our hunt for 'obstacles to understanding,' there's a risk we aim for the wrong target. This isn't about brevity. This isn't about 'short' being good and 'long' being bad; or detail being bad and less detail being good. As we'll see, our pursuit of simplicity is about clarity of language and the removal of distractions and *unnecessary* information. This may mean brevity, but it may not.

It boils down to a question I've asked myself thousands of times – about each sentence of every explanation.

ASK YOURSELF
Is this the simplest way I can say this?

2. ESSENTIAL DETAIL

Detail is good. This might seem to contradict our desire for simplicity, but it doesn't. Allan Little's advice is about simplicity of language, not of subject matter. This isn't about avoiding complex issues or detail. On the contrary, detail is our currency. Detail gives us the facts – the most distilled form of information we have. This is what we're looking to pass on. It's the purpose of us explaining anything.

The reality, though, is that essential detail is not always given the VIP treatment that it deserves.

In the 2010s, I ran up against a counter-argument on detail. An orthodoxy developed in news media that people wouldn't watch long videos. The trend was for videos to be short, snappy and not overburdened with information.

In 2016, I went into YouTube's offices in London. With the help of a range of charts, they showed me how people were watching for longer. For YouTube's users at least, the appetite for detail was there if the subject and video were right. And so, while for some, short videos did (and do . . .) work very well, it was exciting that plenty of people wanted depth too.

Vox in the US realised this was the case more quickly than most media outlets. It launched in 2014 with an emphasis on explanatory journalism and was soon posting videos that ran for far longer than standard TV reports. Soon they were attracting vast audiences – and they still are. And whether it's to an audience of a million or an audience of one, if we're trying to explain something, essential detail is non-negotiable.

I often ask myself: what is the detail I need to include to properly explain this?

If someone is struggling to explain something, often not

enough time has been given to working out what the essential detail is – and then making space for it and explaining *why* it's essential. Whether this is a process of months if you're writing a book (me . . .) or a process of minutes if you're scribbling on a piece of paper before a coffee with a manager (me as well), prioritising essential detail is one of the most important things we can do when explaining.

There are risks here too, though.

Detail for its own sake is not good. It doesn't make us look clever nor is it helpful. Including most of the interesting detail on a subject can be seductive but beware the law of diminishing returns. Every piece of non-essential information makes it harder for the essential information to be communicated.

And this is the challenge for us all as we explain ourselves: can we sift essential detail away from everything else? If we can't, we risk hiding what matters most. If we can – whether you're a plumber explaining the work that needs doing or a manager explaining a restructure or a sports coach running through a change in tactics – *that* is information the people listening want to hear.

If the detail's relevant and helpful, you won't need to persuade people to pay attention.

We'll get into *how* to identify essential information later in the 'Seven-Step Explanation' section.

ASK YOURSELF
What detail is essential to this explanation?

We want to keep it simple and we want essential information. But this essential information won't just include facts. It may need to include complexities that we'd prefer to avoid.

3. COMPLEXITY

The more complicated a subject or an issue, the greater the threat to an effective explanation. If we include complexities and get it wrong, we risk whoever we're addressing feeling confused and overwhelmed or, worse still, offended or ignored. The temptation to walk away from complexity in order to avoid all of that is very real. It also would appear to help our pursuit of simplicity. But just as simplicity doesn't mean ignoring essential detail, nor does it mean ignoring the essential complexities of what we want to explain. Because rare are the subjects that come without complexity. To explain them well, we need to take that on.

EXAMPLE

In the summer of 2015, I got a call from my then editor asking me to go to Greece. It was struggling to meet its national debt repayments and, to put it mildly, there was disagreement over what to do about that. There were protests on the streets, the European Union was working out what help it might offer, the Greek government was working out what terms it might accept in return for that help. For a moment, it seemed possible Greece could drop out of the Euro currency, something that would have had enormous political and economic consequences for the whole of the EU.

For close to a couple of weeks, I spent long baking days on a roof looking down over Syntagma Square in Athens with, to my right, the Greek parliament at the top of it. I was there to do what the BBC calls 'live and continuous'.

This means you are there to present or report whenever BBC News TV output requires. If it's the biggest story in the world – and the Greek debt crisis frequently was that summer – you're required a lot. You may be on air many hours across the day. That's relevant to how you go about explaining the story because there's far too much coverage for you to prepare for each moment. You're not going to know every question you'll be asked or every development you'll need to describe and so you need to be ready for whatever comes your way.

The debt crisis was a story resting on incredibly complex national and international economic and political considerations. The complexity of the issues was matched by their importance. And so, in those first few days, in between applying endless foundation in the forlorn hope of containing the sweat on my forehead, I tried to navigate how to explain those complexities without notes, without huge amounts of prior knowledge, without advance warning of what I'd be asked and without much time on air to do the explaining. If that sounds daunting, it *was* daunting. It *is* daunting when I do similar reporting trips now.

It can be tempting in these moments to gravitate towards the easier parts of the story and to take an 'off-the-shelf' description that someone else has written on the more challenging aspects. But I knew if I did that, I'd come unstuck sooner rather than later. There was nothing for it: I would have to embrace the complexity.

The Athens rooftop which became my temporary home was a little smaller than a tennis court. In one corner we had a gazebo to keep us and the equipment cool between reports. If there was a longer break, there was also a tiny office we could retreat to. It was in there that I drew up three lists.

The first was for the issues I thought I understood. I also wrote how best I thought I could explain them. The next list was for elements of the story I was really struggling to understand, let alone explain. The third was, for the moment, blank. Here I would put other aspects of the story that I didn't understand as I discovered them. (I, correctly, assumed there would be a steady stream of new developments that would flummox me.)

From there, I started to research and to fill in the gaps as best I could.

There was, though, a point that I would inevitably reach where I needed to ask for help. I knew I'd reach this point because I reach it on every story I work on. There's always something I don't understand. And I always ask. I could give you countless examples of the tos-and-fros I perform with colleagues, doctors, teachers, IT support, electricians . . . whoever it might be when I want help with something I don't understand. 'If I said this, would that be right?' I ask.

Earlier in my career, I'd been sent to Brussels to host an hour-long phone-in with the then Secretary General of the Western military alliance, NATO. I'd done my homework but on the morning of the show, I wasn't happy with how I was describing NATO's relationship with Russia. It wasn't that I didn't understand the issue at all – but I wasn't confident how to describe it with clarity and fluency. In these situations, I never ignore my discomfort. On the contrary, I'm looking for it. My inclination to avoid a complexity is a clue that I'm not ready to explain it. In this case, I called my then colleague Jonathan Marcus who indulged me doing the to-and-fro. By the end of the call, and with Jonathan's advice guiding me, I could talk about NATO and Russia with conviction.

This time, in Athens, my mainstay was an exceptional reporter from BBC Business called Joe Lynam. I'd ask Joe about one aspect of the story or another – and then, once he'd explained it, I'd attempt to say it back to him in a simple, short form of words. Often it wouldn't be right, and Joe would patiently correct me. I'd try again. And bit by bit, I'd edge towards an explanation of that issue that made sense to me, would make sense when I used it and was accurate.

The lesson I learned as I watched the drama of Greece's economy and government on the brink was that you can't dodge the complexities and hope to explain something well.

To explain is to first understand.

The reason all this matters is two-fold. First and foremost, a complexity badly explained is, well, complex and inherently hard to understand. That undermines both someone's understanding of what you're trying to convey but also, more broadly, undermines their faith in you as a source of useful information. This then not only impacts the value someone takes from this part of the explanation but there may well be a knock-on impact on your entire effort.

We're all familiar with that moment when someone is explaining something to you and you sense them giving up on trying to make it *all* comprehensible. The subject is so difficult, this part cannot be simplified further, they appear to have concluded. My suspicion in these cases is that the person doing the explanation has either come up against a limit on their own knowledge or their own willingness to make the subject accessible. There may be some good reasons for that – but in terms of explanation this is a problem.

The second reason why engaging in the complexities matters is that you can much better judge what to include. And whether you choose to include the complexity or not, your confidence in your explanation will be greater because you made that decision from a position of strength.

Because that is the twist here. Many times I've spent a long while really working to understand something and I end up not using it. It's still worth doing. It's played its part either way. It gives me confidence in what I do say.

This is both relevant to when you are trying to convey complexities that you understand and those that you're having to learn.

If you're a teacher, you'll already know many techniques to make complex subjects accessible to your students. If you're a doctor, you'll have ways of conveying complex health conditions. My only word of warning to all of us is to constantly check that you're calibrating the explanation to match the level of knowledge of your audience. We can all think of doctors who do this brilliantly and those who leave us lost unless we ask for a more basic explanation.

Or think about this from the other side. Say you're having some work done on your house and you are the go-between as different tradespeople come and go. If you're handling information you don't understand, you risk not passing it on accurately or effectively. You might need to get your head into the plumbing or electrical work sufficiently that you can talk to others about it in a credible and clear way.

In other words, the more you consciously engage with complexity, the better you can judge what information to use and how to use it.

Understanding leads to more accurate assessment, which leads to better explanation. This is why the best explanations always include the complexities.

A great explanation will include the essential details and the essential complexities – *and* do so in the simplest language possible. That simple language will help to meet our next target.

4. EFFICIENCY

The impact that phones have had on our ability to give something our undivided attention remains the focus of both academic research and passionate debate. It is, however, undeniable that we're still willing and able to focus for significant periods of time when we want to. Netflix, Amazon, YouTube, the BBC and others have plenty of evidence that we will listen and watch at length. So too do theatres, cinemas and podcast producers.

There are perhaps two intertwined considerations here: how well we can give something our full attention and how willing we are to do so. And there's no doubt that the competition for our attention is fiercer than ever. Technology offers us an ever-present alternative. It also allows us to shift to something new the moment that we lose interest. As I've watched the evolution of digital technology and digital consumption in the last twenty-five years, one of my responses as a journalist and presenter has been to make sure that if people give me their time, they get a lot in return.

This is a challenge I've been wrestling with for several years. I created the format of the news programme, *Outside Source*, in part to take this on. It was designed to have as much information as possible distilled into each minute. The viewer would be given what I call 'high-protein news' in exchange for their time. Meaning that for every minute of your time that you give me, I'll give you as much usable and useful information in return.

EXAMPLE

During the 2017 UK election campaign, the BBC decided to send me to my home county, Cornwall. Despite being from there, I needed much more than my general knowledge to properly report on the campaign there and so I was on the lookout for more detail.

A night or two before my first broadcasts, one of the main UK news bulletins had a report looking at the town of Penzance. 'Great,' I thought. 'A report not just on the election, or even on the south-west of England, but specifically on Cornwall and the election.'

I watched the report, notepad at the ready. I wound it back and watched it a second time. My pad remained empty. I'd been shown what the place looked like, I heard the views of a couple of locals, some broad information about the area, but in terms of solid, usable pieces of information – statistics, context, analysis – it was slim pickings. It contained some detail, but very little *essential* detail. It wasn't a reward for the time I'd invested. As a consumer, this was a hugely inefficient way for me to learn about the campaign in Cornwall. It sharpened my thinking about the centrality of efficiency to explanation. Because if you're efficient, you give people more in return for their time and they're likely to be glad of that.

As well as giving people 'more for their minute', efficiency provides something else – space. Let me explain what I mean with a short detour via Florida.

In October 2011, I'd arrived in Miami from making a documentary in the Caribbean. (To avoid giving the wrong impression, this is the one and only time this has happened.) I was there to host a series of programmes in Fort Myers, on southwest Florida's Gulf Coast. So I hired a car and headed out into the night. Interstate-75 is long, straight and flat and, at night, there's nothing to see and no one for company. I switched on the local radio news: the death of Steve Jobs, the founder of Apple, was the lead story. It had been announced a few hours earlier.

That night's drive, I heard a raft of tributes and anecdotes about Steve Jobs and one stayed with me: the fish tank story. Here it is as told by former Apple employee Amit Chaudhary.

> When engineers working on the very first iPod completed the prototype, they presented their work to Steve Jobs for his approval. Jobs played with the device, scrutinized it, weighed it in his hands, and promptly rejected it. It was too big.
>
> The engineers explained that it was simply impossible to make it any smaller. Jobs was quiet for a moment. Finally he stood, walked over to an aquarium, and dropped the iPod in the tank. After it touched bottom, bubbles floated to the top.
>
> 'Those are air bubbles,' he snapped. 'That means there's space in there. Make it smaller.'[5]

You can also find a reference to this story in an article in *The Atlantic* from the day after Steve Jobs died. The headline is 'In Praise of Bad Steve'.[6]

I'm not here to assess 'Bad Steve' or indeed any of the Steves he may have been. But one phrase lodged in my mind: 'there's space in there'. And so it is with explanation. There's almost always 'space in there'.

This is something news reporters learn the hard way. 'It needs to be shorter,' we're told by the editor. 'It can't be,' we plead. 'It has to be,' they tell us. Fighting for space is built in to the experience of being a journalist. The front page only has so much space, the news bulletin is only so long and so on. These constraints are non-negotiable. It's the same when you have a fixed duration for a presentation, meeting or appointment. What I'm interested in is how we respond to these constraints.

The risk is that in our effort to operate within them we fall into three traps – we rush our delivery; we cram in too much information; or we jettison valuable information. All these outcomes can harm our explanation.

A couple of years ago, the *New York Times* journalist Jane Bradley tweeted: 'The best editors send you back your copy with 1,000 words lopped off and you don't even notice them missing'. I immediately saved it because Jane is quite right. Fewer words are not a negative and won't automatically reduce the quality of information that you're passing on – often it's the direct opposite.

If I am working on an explanation, I'll do multiple sweeps of what I'm planning to say: repeat drops into the fish tank, if you like. Where can I find the 'bubbles'? Can I find the words or information that are non-essential or that could be replaced with something shorter or simpler? Which sentence is not serving a clear purpose?

I call this process 'tightening'. Later, I'll show you how I do this. Suffice to say, getting an explanation to a point where it is as efficient as it can be is rarely a single act – it's a series of sweeps. The more we do, the more efficient our explanation becomes.

One final point on this: as I mentioned earlier, efficient explanation isn't the same as a brief explanation. It's about maximising the time available to you. When we work on our explainer videos, our concern is that the explanation is efficient rather than it being a fixed length.

If you can be efficient in your explanation, you can both improve the clarity with which you explain yourself and ensure you have as much space as possible to do this.

ASK YOURSELF
Is this the most succinct way I can say this?

We've looked at why we need to keep our explanations efficient. Our next attribute is related: it's about how we avoid confusion and ambiguity.

5. PRECISION

Let's go back to Allan Little and that training course sitting in the BBC archives.

'Good writing is all about choosing the right words to say precisely what you mean,' Allan tells us.

On first reading this may seem so obvious it doesn't need to be said. 'Of course we choose the words we need to say what we mean,' we might think. 'That's what speaking is.' In fact, Allan is on to one of the most common mistakes we all make when trying to explain ourselves – we often *don't say* exactly what we mean.

Getting this right is a two-part process. The first is working out what we want to say.

As Allan Little said to me recently, 'If your sentences are too long, your writing hasn't been disciplined enough. If your writing hasn't been disciplined enough, your thinking hasn't been disciplined enough.'

My most ineffective efforts are always when I've not worked this through – with that visit to Canary Wharf being top of my list of case studies. There are many others. When I feel myself doing better, I'm clearer on what I'm trying to explain overall and what purpose each sentence is serving.

But even when we know what we want to say and we know the role each sentence, paragraph or section is serving – that's not the same as 'choosing the right words to say precisely what you mean.'

I am constantly asking myself – are my words conveying what I want to convey? If the answer is no, then I change them until they do. To do this requires 'precision' and, not for the first time, I've found inspiration for how to do this when I was least expecting it.

While Steve Jobs was dropping iPods in fish tanks in Palo Alto, further south many years earlier, Joni Mitchell recorded her album *Blue* in Los Angeles in the early seventies.

There aren't many albums I've listened to more than *Blue*. I found it in my college's record library and have proceeded to put it on relentlessly ever since. There is something incredibly direct and arresting about the entire album.

There's an honesty to the lyrics that make them almost too intimate to hear. There's a beauty and virtuosity to the melodies and performance too. When they are all combined, Joni Mitchell reaches somewhere remarkable and new. But despite listening to the album hundreds if not thousands of times, I'd never spent any time considering *how* Joni Mitchell achieves an alchemy that others armed with a piano, a guitar and a desire to share have not. What is she doing that explains to me so clearly how she feels and, magically, helps me understand how I feel too?

Having heard me play it any number of times, not long ago my parents gave me David Yaffe's breathtaking biography *Reckless Daughter: A Portrait of Joni Mitchell*. Like all the best critics from music to sport to art – David Yaffe helps you see afresh something you've encountered countless times.

In his book, Yaffe writes: 'The sounds are acoustic and spare. Drums are brushed and muted. Guitars and other string instruments – including that coveted Studio C grand piano – are acoustic. The emotions are out on the surface, beating and exposed like Russ Kunkel's conga.'

'Spare.' This was a lesson right there for our own explanations. To me, David Yaffe was saying that the content of what Joni Mitchell had to offer was so potent, she had no need to overplay the arrangement. If we adjust that for explanation, if what we have to offer is sufficiently relevant, interesting and helpful, the words around that information can be 'spare' or sparse too. Less is more in explanation if the information at the heart of it is worth hearing.

This was one lesson. Another came when, inspired by David Yaffe's book, I started reading more about *Blue*.

I found a list that NPR in the US has produced which runs through the '150 Greatest Albums Made by Women'. *Blue* is number one. In the text explaining the decision, Ann Powers of NPR Music writes, '*Blue* reminds us that emotional writing is only powerful when it is punishingly precise.'

There was that word again. 'Precise.' 'Punishingly precise', no less. I knew exactly what Ann Powers meant.

The more I thought about the role of precision in explanation and communication of all types, the more I saw it. It demands that we choose just the right words and that we clear away anything around them that might distract.

There are of course many ways to convey meaning – whether in art, music, journalism, conversations or events. Precision is not a pre-requisite to communication.

However, if high-impact explanation is our goal, I'd argue precision is essential.

ASK YOURSELF
Am I saying exactly what I want to communicate?

Having established that we're wanting simplicity, essential information, complexity, efficiency and precision from our explanation, the next ambition is to make sure that, if all those things are achieved, we also explain why what we're saying matters.

6. CONTEXT

Context is everything in explanation. Nothing in the human experience exists in a vacuum. Something matters because of its relation to other events, people or knowledge. We all know this instinctively, but often when we explain something, context is jettisoned in exchange for more detail on the immediate subject or event itself. In a news situation, that would mean spending all your time on the event that has just happened and hardly any on the circumstances that led up to it. If you do that, you both risk people not caring as much as they might about what you're describing and them not understanding it as well as they could. Unless we explain the context, we can't assume the people we're speaking to will understand why this particular subject matters. And without that understanding, any form of explanation will struggle to be effective.

At the dentist, if your tooth hurts, it's relevant that it's hurt for over a week.

If you're telling an interviewer that you achieved sales of a certain level, it's relevant what they were before you took over.

If you're making the case for a change to how your team is structured at work, it's relevant the structure's not been changed for ten years despite several requests.

Context is key to making people care and making them understand.

EXAMPLE

One of the greatest demonstrations of this came when I was presenting on the BBC World Service in 2011. News had come through that the Governor of Punjab Province in Pakistan, Salman Taseer, had been assassinated by one of his bodyguards. This was a huge story in Pakistan.

Salman Taseer was a high-profile figure there. He wasn't, though, very well known around the world.

Also in the newsroom on this day was Owen Bennett-Jones, one of our most senior presenters and a man steeped in Pakistan's past and present. We asked if he'd talk to us and, off the back of the news summary, I read a brief introduction and introduced Owen.

What followed was a five-minute spell of the best explanatory journalism I've ever heard. With no notes, Owen proceeded to unpack why this mattered so much not just to Pakistan but to all of us. After all, the vast majority of people listening were not in Pakistan. They were in Oregon and Lagos and Sydney and Brussels. Most, like me, would have been coming to the story with little previous knowledge.

Of that five minutes, less than sixty seconds were spent on what we knew of the attack, the rest was spent on context. Owen placed Salman Taseer's career within the context of Pakistan's democracy, of the intense divisions around Pakistan's blasphemy laws, the relationship between Pakistan and the West, and the delicate security situation in Pakistan and the region.

There are times when a piece of live broadcasting is so captivating you can almost feel the world slowing to see or hear it. These are moments that make you stay in the car at the supermarket or mean you're late leaving the house. I didn't say anything, my colleagues in the gallery didn't say anything. We all just listened.

This is in part a tribute to Owen's eloquence and knowledge. But it was also a masterclass in understanding the centrality of context. Owen knew that to convey the importance of this moment to people listening around the world, he needed to spend time on the background to the assassination. In doing so he not only taught me about Pakistan but also about the art of explaining.

In the years since, I've given more and more space to context. If you're a regular viewer or listener of mine, these are some phrases you may well recognise.

- 'The reason this matters is . . .'
- 'To understand this, we need to remember that . . .'
- 'All of which connects back to . . .'
- 'This is not happening in isolation.'
- 'This is important beyond the immediate consequences because . . .'

You get the idea. Too often in the past, I have found myself announcing that US interest rates were staying the same or that the World Bank had a new president or that one election or another was happening without offering any help to viewers and listeners who might be asking, 'Why does this matter?'

Owen's monologue on Salman Taseer taught me the importance of answering that question directly – not just when I'm broadcasting but when I'm explaining myself in any scenario. I want to tell people explicitly – this is why this matters.

There are multiple benefits to doing this well.

First, you'll significantly increase your chances of someone wanting to hear what you say. Then, if they do listen, their understanding will be enhanced by the context you provide.

There is a reverse benefit of trying to provide context. If you're struggling to lay out why what you're planning to say matters – or why one part of what you have to say matters – this should guide how much emphasis and time you give to it. Not everything matters enough to mention and, if you don't feel it matters that much to you, you can be absolutely sure that the people you are addressing will sense that. Many is the time in the newsroom we'll be unsure whether to keep a section of a story or a video. 'It's interesting but is it essential?' we'll ask. Context can help you work that out.

ASK YOURSELF
Why does this matter to the people I'm addressing?

Now, if we've addressed all six of our criteria so far, then our explanation will be shaping up well. But we need to be sure that within it we're not doing things that create distractions. Allan Little's emphasis on simplicity of language and pursuit of those 'obstacles to comprehension' helps a great deal with that. But there's still more that can be done.

7. NO DISTRACTIONS

One of the great threats to all the good work that may have been done if we follow points 1-6 comes from our own words and actions. Explanations can be undone by distractions of our own making. It might seem a strange thing to say, but a good explanation requires the person or people we're communicating with to stay focused on the subject in hand. All of us are remarkably good at including any number of things that may disrupt that focus.

These distractions can come in two forms – verbal and visual. Let's take them in turn.

We can all think of a time when we've been in a conversation and someone has introduced a word or a name that we don't know.

'I said to Sam that he won't be able to do that.' But we don't know who Sam is.

'Sam has been angry ever since the project review.' But we don't know what the project review is.

'All Sam is doing is casuistical evasions.' But we don't know what 'casuistical' means.

(I do now know what it means as I've just read it in a newspaper editorial and had to look it up. It's the adjective of casuist. I'm told by Oxford Languages Google dictionary that this is 'a person who uses clever but unsound reasoning, especially in relation to moral questions; a sophist.' So now I know. But I didn't when I was reading the editorial and so I didn't understand the sentence.)

Sometimes when this happens, we can make an educated guess at who the person or event that's been referred to is or what the word we don't understand means. But it's disconcerting. At the very least, it makes it harder to understand what the other person is telling you. But you also know that they're not taking care to include you in what they're saying. They either didn't think to, don't care to or have misjudged your level of knowledge. Either way, that can jar.

To stop it from happening, we need to think about why we do it. Creating confusion isn't something we'd set out to do. But one of the reasons we end up there is that, sometimes, not explaining something can be tempting.

I think of this as my NATO rule. NATO comes up in many stories I cover – I mentioned it earlier. I must use its name on air hundreds of times a year. NATO is the North Atlantic Treaty Organisation. It's a military alliance that was created in 1949 by the US, Canada and a number of Western countries including the UK. Since then it's expanded and recently Finland and Sweden formally began the process of becoming members. NATO is central to many of the biggest stories. To take two examples, if you want to understand the West's withdrawal from Afghanistan in 2021 or if you want to understand the West's response to Russia's invasion of Ukraine in 2022, you need to consider NATO.

Reading that last paragraph, some of you will know all that and a lot more, and some of you may have learned something. We all have different degrees of knowledge across every subject. How we judge how much the person we're addressing knows is crucial to the success of an explanation.

Now, if I were referencing NATO in a talk or video and I was confident everyone in the audience knew what it was, that'd be the end of the matter. I could concentrate on explaining something else.

But what about when you can't be sure what everyone knows? To continue with this example, what happens if I reference NATO without explaining it?

If I were to do this – to include a reference that not everyone understands – I can expect a number of things to follow and none of them are good.

The moment a person hears a reference to something they either don't know at all or aren't sure about, they'll immediately start wondering – or trying to remember – what they know about it. If that happens, then the first problem is that they don't understand what you're referencing or why. Worse, though, is that while they're pondering, they're not going to be focused on whatever it is you're saying next. And, third, if you're referencing something they don't know without explanation, they may, quite reasonably, conclude that what you're saying is not aimed at or relevant to them. And all of this increases the chances of them tuning out.

For all these reasons, I am strict about explaining a reference or getting rid of it. I'm under no illusions that if I keep it, that's going to take a sentence and, sometimes, more. If I'm under severe time pressure, I may struggle to do that. In which case I then move to decide whether the reference is necessary.

These are fine and difficult judgements. As I'm writing this, I'm thinking of many pained discussions with colleagues about whether a reference stays or goes. We're going to explore how to make those judgements later. But as a starting point, let's take it as read that leaving in a reference that can confuse or distract isn't an option. That's my 'NATO rule'. Here's one example of this being judged perfectly.

EXAMPLE

While writing this book, I've been listening to the BBC's podcast *The Lazarus Heist*. It's about North Korean cyber-crime and in one episode it details an audacious hack on the Bank of Bangladesh that almost results in a billion dollars being stolen.

The journalist Geoff White is the co-host and he describes how investigators who've been called in by the bank 'start to focus on the computers themselves. And one set in particular. The machines that run a system called Swift.'

The moment I heard the word 'Swift', I started thinking, 'What's that?'

Moments later, White continues, 'We need to take a minute to explain what Swift is because it's a huge part of the story.'

What follows is exemplary. The pause in the overall story doesn't undermine the telling because the explanation is fascinating. It's interesting in its own right (in fact it made me want to know a lot more about Swift), it sets me up to understand the story even better and it makes me want to hear the story even more. Far from undermining my engagement in the story, the pause for this specific explanation deepens my engagement and understanding.

How to judge what to explain and what not to is vital because every word has the potential to demand something of those we're addressing.

To help me with this, I like to split the words that I use into four categories. Here they are with a handful of examples which I've picked at random, imagining I am talking to a global audience as I do on the news.

1. **Support words** These are words like 'the", 'it', 'then', 'of', etc. The brain expends next to no energy processing these and they require no explanation.

2. **Known words** This can be anything from a noun like 'table', to a country like the United States, to a process like

an election, to an event like the Second World War. The vast majority of people will expend minimal energy on these because they are very well known to us.

3. **Partially known words** This could be anything from a concept like 'inflation' to a system like 'proportional representation' to an agreement such as the 'Paris Climate Accord'. Some of those listening definitely won't understand these references.

4. **Unknown words** This could be anything I can't reasonably expect the vast majority of people listening to know. 'Proroguing' parliament, for example.

Which words are in which group is dictated by assessing who you are talking to. And primarily, I'd want to be thinking about which words fall into the last two categories. Then I can decide whether to explain them or delete them.

In the case of partially known words, I often use the phrase 'as some of you may know . . .' It acknowledges to some people that you're telling them information they don't need, while informing those who do need it.

Assessing the language that you're using is crucial to removing distractions in your explanations. It's often not done because it's less work if you don't bother to explain words that are partially unknown or unknown. But time spent doing this is time well spent. Just as any temptations to use clever or complex vocabulary to look, well, clever or complex should be resisted, so should making references for the same purpose. They'll distract and hinder what you're trying to do.

If the 'NATO rule' addresses verbal distractions, what I call my '*Outside Source* rule' addresses visual distractions.

As I was explaining earlier, I wanted the *Outside Source* format to allow me as the presenter to construct the story in front of the viewer. To do this effectively, I was adamant that each time they saw something it'd be *directly* connected to the aspect of the story I was describing at that point. Too often when I'd been watching TV news, pictures were being used as wallpaper. If the reporter was talking about a war, there were some general pictures of the fighting, or if they were talking about transport policy there might be some pictures of buses or trains. To me this produced three undesirable outcomes – it told me that there was nothing specifically relevant to see and so decreased my engagement; it distracted me rather than engaged me; and it created a sense of a lack of purpose. I didn't want any of those outcomes when I was trying to explain a news story. Nor do we want these outcomes in any explanation we're giving that comes with a visual accompaniment.

I am still strict on this approach in all my presentations, videos and reporting. I only show pictures, graphics, maps or tweets if I explicitly reference them. And we only show them *when* I explicitly reference them. This greatly supports my explanation. If they are there when I'm not talking about them, it does the reverse.

In my experience, we take a chance if we hope the people receiving our explanation are good at handling two competing sources of information. This is not to pass judgement. It's something we all find hard. We can move quickly between a number of information sources, but we can struggle to process them *simultaneously*.

When I'm on TV, in my earpiece I can hear both the programme as well as the director and the editor in the gallery. It took a long while to get used to that, it remains hard and, as my colleagues will tell you, I still miss things. Now, that is how it needs to be in a TV presenter's ear, but when we're doing explanations – on TV or anywhere else – we don't need to offer multiple competing information sources. Or, to put this more strongly, that is the last thing we should be doing when we're trying to communicate clearly. Not for the first time, we'd be undermining our own endeavours.

I'm hoping, as you read this, you're starting to think of times you've seen this done. How often have you been listening to a presentation at work or at a conference or watching a YouTube video and slides are coming up with lots and lots of information on them?

At best, they show you what the speaker is going to say next. At worst, these are images or graphics that don't yet connect to anything that is being said. Either way, quite understandably, our mind wanders on to them. 'What is that?' we think. And, in that moment, we stop listening properly.

I came across this recently when interviewing someone for a job. They were a great candidate and full of ideas. The only problem was that they had put a lot of their ideas on a single slide of their presentation. The moment it flashed up, there was a huge amount of information on the screen and a lot of it looked interesting! I had to work to stop myself from reading it all. If I had done, I'd have stopped listening properly.

On another occasion I once sent a sixty-slide PowerPoint presentation in advance of me appearing at a conference. 'You only have ten minutes, you know?' was the gist of the reply. Ten minutes was going to be fine. I didn't have a particularly large amount of information – I had just put each piece of supporting visual information on its own. It was there when I needed it and not distracting the audience when I didn't.

Visual support can be a massive plus in great explanation. But if you're creating visual distractions, you'll be having the opposite of the desired effect.

Hunting down distractions in all their forms is essential to giving your explanation the best chance to communicate what you want to say.

Good explanation minimises distraction, maximises context and gives you the most important information with precision, simplicity and efficiency. And as I try to achieve that, three more guiding principles can help me. The first is as obvious as it is important: if someone's not listening to you, you're not going to be communicating with them.

8. ENGAGING

In 2002, the BBC moved me to a 5 Live show called *Up All Night*. We were a small band of producers tasked with filling the darkest hours from one until five in the morning. Not unreasonably, we didn't always receive as much attention from the bosses as some of the other shows, but on one occasion we were asked to attend an 'audience briefing' in the controller's offices.

These rooms were on a different floor to where we worked and were where the big decisions got made.

Back in 2002, I was a junior producer and so I rarely went to the controller's floor, which is why I can remember so clearly where this briefing happened.

We sat down on the sofas and for an hour were given a whole raft of details on how and why people were listening. All these years on, I don't remember all of that now, but one idea lodged in my mind and it's never left.

At the time, the BBC was doing audience research using a system of live listening. It would bring together a group of people who'd listen to a programme as it was being broadcast. Each person would have a dial that they'd hold. The dial would start on 0 and, as they listened, the participants would turn the dial right if they were enjoying the programme, left if they weren't. The further they turned right or left, the more they liked or disliked that moment. After an hour of doing that, the researchers would have a detailed account of how each individual listener and the group as a whole felt about the programme. Similar technology is used to produce 'the worm' that you sometimes see on screen during leaders' TV debates ahead of elections. Each candidate has a worm that rises or falls on the approval scale as the focus group responds to what they're seeing.

Now, both in the context of radio audience research and election programmes, there's a broader discussion about how accurately this kind of research captures actual behaviour and feelings. For me, though, the power in that presentation back in 2002 was in what it represented.

What we could clearly see in the data was that there were moments of weakness in any given hour of programming. Some of the moments are familiar to broadcasters. The end of half hours and hours can be switch-off moments, just as the 'top' of the hour is a switch-on moment. What was revelatory to me was the consistency with which listeners were losing interest at other points. This wasn't just happening between items; it was happening *in the middle* of them. And so started an ongoing preoccupation with trying to understand the reasons why that happens and what can be done to avoid an imaginary dial drifting to the left.

This research was specifically looking at radio consumption. But the importance of what I call the 'dial test' holds regardless of what you're doing. It is self-evidently true that if someone is not listening to you, you're not going to explain anything to them.

If we look more broadly at the digital world, people move on ruthlessly – and we have the data to show us exactly how. If a video, article, programme or podcast isn't keeping someone's attention for a moment, they are gone, and we can measure when.

This creates a pressure, but it also ups your game. Where are the moments when I might lose someone? Is there part of what I plan to say that lacks clarity or purpose? What are the reasons I am giving someone to keep going with what I have to say?

This applies to any form of explanation – whether it's a video, presentation, speech, briefing note, article or any number of other examples.

I think in terms of the dial test all the time. It's become a byword for trying to maintain a consistent level and to spot a moment of weakness in how I'm trying to explain myself.

EXAMPLE

In 2021, I posted a video about Donald Trump's deal with the Taliban in 2020 and how it connected to the decisions that Joe Biden made around the US withdrawal from Afghanistan. It racked up over a million views on Twitter. But that impact could easily not have happened.

After recording one version, my editor on the day, Andrew Bryson, intervened. He kept coming back to one section in the middle that he felt was losing us momentum. Along with producer Tom Brada, we went back over it several times, removing, adding and moving around various elements. Our concern was not the overall length of the video, it was the 'dial test'. Was it too wordy? Was it clear what purpose this section was serving? Was the information essential or simply interesting? Were we giving a reason for people's attention to drift? Or even switch off?

I was wedded to some of the detail in this section. Andrew insisted it didn't keep the viewer, wasn't laid out well enough and wasn't essential. We watched back my longer version once more. Andrew was right. He'd spotted the moment of weakness and Andrew's version was the one I posted. The information we'd discarded wasn't essential and, as the video went viral, there was vindication that Andrew had made sure it passed the dial test.

If we fail to look for these moments, we're taking a huge chance. If you lose people in one section, they might not see or hear anything else that you have to say. Even if you're giving a speech, and have a captive audience, losing people in one section is still the last thing you want. There's no guarantee their attention will return once your explanation returns to form. Or if you're a teacher and your explanation loses the attention of some students, that can quickly lead to all the kids getting restless. The best explanations connect each section, part by part. But if one link in the chain underperforms, that can impact the whole thing.

The dial test is a constant reminder that to explain ourselves well, we need to remain acutely aware of whether we're keeping the attention of the people we're speaking to.

ASK YOURSELF
Are there moments when attention could waver?

There are any number of techniques that can be used to keep someone's attention, and one of the most effective is something any good piece of explanation does – it explicitly answers the questions that the audience has.

9. USEFUL

The best explanations are helpful. Whenever I want to explain something, I write a list of the questions that I think I'll be expected to answer. If you can answer them all, there's a good chance whoever you're addressing will want to hear what you say.

EXAMPLE

In late 2019 and early 2020, Australia was tackling a series of huge bushfires. At the time, one of our most experienced producers was former ABC reporter Courtney Bembridge. Courtney was keen to really take on the story and across several weeks we posted a series of videos. The subjects we chose were often guided by the questions we could see that people had about the fires and the response to them. Questions like:

- Did Prime Minister Scott Morrison's statements on climate change match what others in his party were saying?
- Was a failure to do more 'hazard reduction burning' a factor?
- Was arson responsible for the fires?
- Was the coal industry impacting Australia's response to the bushfires?

On the list went. Video after video explicitly sought to answer the questions we could see people asking. Video after video was heavily shared.

Whatever the type of explanation you're giving, if you are catering to a need or desire for particular information, this will serve you and whoever you're speaking to well.

From a meeting about some prospective work, to a conversation with a music teacher about a band you want to be in, to a presentation to the local council on an issue you care about – if you can anticipate the information that people want from you, then you're much more likely to deliver it in a coherent and focused way. It also increases your credibility and increases their engagement.

ASK YOURSELF
Have I answered the questions that people have?

Remaining focused on keeping people's attention and answering their questions are both great foundations for an effective explanation. But there's still one more possibility of things going awry. The final attribute of quality explanation is one that ensures all the previous nine aspirations are working towards one outcome.

10. CLARITY OF PURPOSE

Something I've learned the hard way is that if you're not sure exactly what you're trying to do or say, people tend to notice. That was certainly the case in my visit to the offices of the *Independent* and I can easily recall many moments that have taught me this lesson. It's not a great feeling and, in those moments, it can be very hard to regain momentum and the attention of those you're addressing. My main strategy for avoiding this is to address the problem in advance.

EXAMPLE

One lesson on clarity of purpose comes a long way from the world of journalism and TV – in this case, from DJing.

Being a teenager in the early nineties meant that, like many others, I got caught up in dance music. What began as avid consumption of cassettes of live sets morphed into a vinyl habit and, by my final year at university, I had two turntables and a mixer in my room, a pile of records against my chest of drawers, and I'd started to DJ – though admittedly mostly in my bedroom at first.

Ten years later I was booked to DJ at an event in Regent's Park with a big stage for bands, the usual array of veggie burger and crêpe outlets and a tent with a sound system.

I had no real feel for what the crowd would be or what kind of music they'd be expecting. I still have the CDs I made of the tracks that I chose. They're all great tunes – but together they didn't make too much sense. I'd pull people in one direction with a run of two or three tunes and then lose them a little with the next one. None of this mattered in the scheme of things. It was still a good time. But the feeling that day of not being sure what I wanted to do stayed with me. Just as it has at some other gigs when I've not properly worked out what I want to play. Equally, the feeling of DJing when you manage to sync with what the crowd wants to hear is exhilarating.

A few days after the Regent's Park event, our eldest daughter, Alice, was born and that was pretty much that for my DJing days (until a recent and improbable return to it). But the link between these DJing lessons and my efforts to explain is useful to me. That feeling of a chance opening up and of not being quite sure what to do with it is familiar to me from occasions on the turntables. These are moments where I'm not clear on the purpose I'm working towards. And as I've focused more and more on explanation as a discipline, I've understood that clarity of purpose is what helps us communicate – whether it's with music or words. Without it, I easily end up with that same feeling from when I was playing records in Regent's Park – of having the raw materials but not knowing quite what I want to say.

This thinking is now hardwired into my approach to telling the news. All journalists are familiar with the two questions – what's the lead? And what's the top line? Meaning *What is our most important story?* And *What is the first thing we want to say about that story?* When I think of the purpose of an explanation, though, I'm thinking beyond those calculations, important as they are. I'm thinking of what the entirety of my telling of this story is trying to achieve. I want that purpose to run through every word.

This thought applies to hundreds of different types of explanation and communication.

If you're writing an essay at school, the precise question you're answering should guide every decision you make about what to include.

If you're giving a work briefing about a new budget and how it will impact colleagues, if you've information that isn't explicitly helping to explain the nature of the budget or of the changes that will follow, it stays out.

If you're raising money for a charity and hosting a local event, everything you say needs to work towards the goal of people feeling motivated to donate.

This may seem self-evident, but we habitually fill our explanations with information that, while perfectly interesting, isn't supporting our overall purpose. And when we do this, our explanations inevitably have, well, less purpose.

One of the most powerful tests I use when I'm reviewing any piece of communication or explanation is to go through it step by step and ask of everything I'm planning to say – *is it explicitly supporting the overall purpose?* If you can remove anything which isn't doing that, you end up with a collection of information that is completely aligned. This is a perfect complement to the work you've also done on making sure each piece of information is presented with clarity and simplicity.

The purpose becomes the guide to every decision I take when preparing an explanation.

This is why it's rare a week will go by when I don't say to a colleague, 'But what are we actually trying to say here?' There may be nothing wrong with the elements of a story we've assembled but the way my script reads, the sense of purpose is missing. Or there may be multiple purposes pulling in different directions. We're trying to do several things and risking doing none well. Get that wrong and your chances of taking someone with you reduces significantly. Get it right and everything connects to the single purpose. The overall effect can be potent and deliver high-impact communication.

> **ASK YOURSELF**
> Above all else, what am I trying to explain?

THE ANATOMY OF A GOOD EXPLANATION: IN SHORT

This leaves us with ten vital questions to keep in mind whenever we're explaining ourselves. It might seem a little over the top to check all ten every time you set about an explanation, but these questions quickly become habits.

At first, you ask them consciously. In time, they simply become part of how you approach communication in all its forms. And while I do much of this on instinct now, I still use these questions. They are incredibly effective at keeping me on track.

Rather like playing a sport, being good at explaining is about learning skills but also about being sure to keep sharpening them. I still keep my lists of things to check to hand and hopefully this is one you can use too.

1. SIMPLICITY
Is this the simplest way I can say this?

2. ESSENTIAL DETAIL
What detail is essential to this explanation?

3. COMPLEXITY
Are there elements of this subject I don't understand?

4. EFFICIENCY
Is this the most succinct way I can say this?

5. PRECISION
Am I saying exactly what I want to communicate?

6. CONTEXT
Why does this matter to the people I'm addressing?

7. NO DISTRACTIONS
Are there verbal, written or visual distractions?

8. ENGAGING
Are there moments when attention could waver?

9. USEFUL
Have I answered the questions that people have?

10. CLARITY OF PURPOSE
Above all else, what am I trying to explain?

Now that we've considered the ten attributes of a good explanation, we need to consider who we're doing this for.

2

KNOW YOUR AUDIENCE

If you run a business, you'll know who your customers are. If you make a TV programme, you'll know who your target audience is. If you're making a speech to an industry conference, you'll know who is attending. There are many times when we are conscious and clear about which people we're communicating with. We factor that into what we make and what we say. In my experience, we don't do this in many other areas of our lives. We don't shape and adapt our explanations according to who they are for. Part of the art of explanation is to do just that in whatever circumstances we have to explain ourselves. In the same way that you can't run a business as well as possible without under-standing your target market, you can't explain yourself as well as you'd like if you've not considered who your explanation is for.

As often as possible when communicating, I want to pause and check what I know about the people I'm speaking to or writing for. If I'm not satisfied that I know enough, I'd try to find out more by asking or researching. For me, the equation is very simple. The more you know about who you are speaking to, the more you can calibrate your explanation and the more likely you are to communicate effectively. Let's unpack how we can do that for anything from a set-piece job interview to a spontaneous or unexpected conversation with a teacher at school or a lecturer at university.

These five questions are a useful start:

1. The target: who am I talking to?

I always want to get as much information as I can before doing any explanation. If I'm speaking at a conference, who are the delegates? If I'm at a school, how many children will there be? What is the age range? Are there particular subjects they major in? If I'm at a company, which staff are coming and what do their jobs entail? If I'm writing an email, who's getting it and why? If I've got a hospital appointment, which doctors and nurses will be there, and what are their respective areas of expertise? Who will take decisions about my treatment and who won't?

As with so much of the advice in this book, you can and should adapt it to fit your circumstances. If I was launching a new journalism product for a global audience, I'd spend weeks if not months understanding who I wanted to explain the product to (that could be colleagues, distributors or the audience itself). That understanding could make or break the product's success. If I am called into a meeting at five minutes' notice with a senior figure at the BBC, I still stop to think about who they are, how they fit into the organisation, why they might want to hear from me and what information I might want from them. I could do a form of that in sixty seconds. It won't be perfect, but it's a lot better than nothing (see page 280 in 'Quick Explanations' for more detail on this).

Whatever the circumstances, being conscious of who we are offering our explanation to puts us in a better position.

2. Knowledge assessment: on this subject, what do they know and what would they like to know?

Of course, you aren't going to know the totality of anyone's knowledge nor the gaps within it. However, we can make informed assessments of the level of knowledge within our audience. This gets harder the larger and the more diverse that audience becomes but it shouldn't stop us trying. Even if the conclusion is that we know very little about who we're communicating with, it's far better to go into that scenario fully aware of that.

There will be some scenarios – for instance a talk to a group of colleagues who do the same job as you – where you can accurately assess the level of knowledge. There will be others – such as a presentation at a conference that is open to the public – where you know next to nothing about the people you're meeting or addressing.

In these situations, I would ask questions in the meeting or presentation to help me better understand who they are. I can calibrate what I say according to what I'm told. People will rarely mind you doing this. You're taking an interest in them and in making sure what you say is relevant to them. That's not going to cause offence. What can cause offence is if the information you're asking for is easily available. Then you'll look like you've not made an effort.

Thinking about what your audience knows is important. Thinking about what they'd like to know is vital too. We looked at that earlier in the case of my bushfires videos. But if I look back through the mists of time to a stint as a contributor on British Airways inflight radio, I definitely didn't ask myself this question.

This was in the bygone era before smartphones, when if you didn't fancy the one film that was showing on the plane, you could plug in the headphones that came with the small tube of toothpaste and the eye mask, and tune into a range of 'stations' which were playing on a loop. One of them was about culture. It was presented by the formidable John Wilson, who's still on BBC Radio 4. The idea was that I'd pick out three cultural events around the world and we'd talk about them for fifteen minutes. On the face of it, this was a simple proposition: pick the events, talk to John.

For each edition, I would assemble and organise a vast amount of information about these events. But who was listening to me? People on a plane. Only people on a plane. And were these people really listening for a detailed practical guide to what was happening at a particular event? If I'd thought about it, the answer would have been absolutely not. Most of the listeners wouldn't ever be in the country where the event was taking place. The slot just needed to be enjoyable, fun, descriptive and easy to listen to. That was more important than cramming in every last piece of information on the practicalities of the event. Cramming, though, was what I did.

The subjects were incredibly diverse – a Cherry Picking Festival here, an International Piano Festival there, a Bahamian carnival next. I often think John wasn't quite sure what had hit him. One evening, I inflicted my fifteen-minute effort on my friend Joe. 'There's a lot in there, pal,' he noted drily at the end. And, indeed, there was. Far too much. In these early days, I lacked the confidence to leave information out (something we will turn to later). If you were on those flights, I can only apologise if it was a bit much.

If we can better understand what our audience knows and wants, we can better judge what information is most relevant and helpful to them, with all the benefits that come with that.

3. Tailor it: how do they like to receive information?

On a macro scale, the BBC and many other organisations have audience research that can detail how people of different ages and backgrounds consume information and products. The content they produce for one group or another can be shaped by that knowledge.

If these are macro considerations – for organisations and businesses communicating with large numbers of people – you might not think we need to make the same calculations on a personal level, but to give our explanations the best chance, we do.

To not use every available piece of information about how your intended audience likes to receive information is a missed trick. You risk having the right message but offering it in the wrong way.

Here's one example from an important moment in my career.

EXAMPLE

From 2013 to 2018, James Harding was Director of BBC News. He was a high-energy presence in the office – sleeves always rolled up, shirt open one more button than most would go, very much of his own mind. He was a man who engaged with you intensely but whose attention you could easily lose. As one colleague put it, 'James is charging ahead and then looks over his shoulder to see if the BBC has caught him up.' Not always, was the answer to that.

BBC News employs thousands of people and a one-on-one meeting with the Director of News is very rare. However, in 2014, I inquired if I could get a meeting with James because I needed help changing my presenting commitments. I also wanted to evangelise about how I thought we could be streaming new types of news programming.

The meeting was put in the diary. Fifteen minutes. Fair enough, the Director of News is very busy.

I prepared two short presentations on each subject. I practised each. And, as I did, I thought back to previous interactions I'd had with James and how I'd observed him in meetings. He was very quick to pick up points and would briskly move the conversation on if he felt he'd understood what was being said.

Having considered what I knew, I discarded some of what I had and practised my presentations again, each time asking whether what was left was essential.

By the time the meeting came, I had forty-five seconds on my work commitments and ninety seconds on streaming. Towards the end of the ninety, I could feel James wanting to speak!

Within a few weeks a shift in my commitments was agreed as was further support to experiment with streaming.

This was a scenario where I'd really thought through who I was speaking to, what detail was needed and what wasn't, how long I would have and where I was most vulnerable to losing my audience's attention. On this occasion it paid off.

Whether macro or personal, the more we assess how people prefer to consume information, the better we can tailor how we explain ourselves. The better we tailor how we explain ourselves, the better our chance of that explanation engaging our audience. The better the engagement, the better the chance of the explanation being understood.

Already, then, we've considered who we're talking to, what they do and don't know on this subject and how they like information to be provided – next, we want to make sure that our explanation is for them in particular.

4. Make it personal: how best can you convey that this information is for them?

If you think something is for *you*, you are much more likely to pay it attention. We live in a world where a tsunami of information is coming at us from the moment we wake. Inevitably, we are constantly, and often unconsciously, making decisions about what information to give our attention to and which to ignore or, at the least, not properly engage with. This is the context to all of our efforts to communicate. When I met James Harding, he was, for a few minutes at least, a captive audience. But this is the exception. Most of the time, we are constantly competing for people's attention. One of the most effective ways of getting that attention is to make your intended audience realise that what you have is for them.

Let's take email as an example. The volume of emails that many of us receive is a problem. Working through your inbox can feel like a job in and of itself. But while there's a plethora of other ways to communicate with each other, email remains dominant in many aspects of the working world and many of us need to use it.

However, the more emails people receive, the lower the chances of them reading the one from you.

This is an ever-increasing problem for companies. The attention staff can give to each piece of communication is inevitably dropping as the volume increases.

I raise all this because my efforts to get replies to emails do really make a difference to the work I'm doing – and those efforts have shown me one dynamic which applies far beyond our inboxes. I keep it in mind whenever I am explaining myself.

It's this: if someone feels you are talking to them, they are far more likely to engage and respond. Equally, if someone feels you are communicating with a group but not particularly with them, they pay less attention and are less likely to respond.

You may know the old saying about work meetings: 'If this meeting could happen without me, I shouldn't be here'. There's something related to that with group email. If someone feels they could not reply and there would be no particular consequence, they probably won't. They'll decide – consciously or often subconsciously – someone else will pick this up.

Having observed this becoming more and more pronounced, in the late 2010s, I started to think what I could do about it.

I'm aware as I write this that admitting you're someone who thinks about how to better elicit replies to emails doesn't make me sound like great company on a Saturday night. Hear me out! On the face of it, this is really not something to set the blood racing. But for me, while I was trying to get various ideas off the ground, getting replies to email – especially replies that really engaged with the content of the emails – made a huge difference to the success or otherwise of what I was doing.

I decided to try an experiment. Instead of making the subject of the email something like 'our budget' or 'marketing query' or 'plans for the conference next week', instead I wrote the person's name. 'Hi Diana', 'Hi Sachin', 'Hi Andrew', and so on. More often than not, it worked. Before the person had even opened the email, they knew it was from me and it was for them. I'll talk more broadly about short written communication on page 282. For now, though, the lesson here is to always think about how to make it clear your explanation is personally tailored for its recipient.

EXAMPLE

Making people feel that you're communicating with them gets a lot harder the more people there are. My approach to this is rooted in a technique that I learned while presenting a phone-in on BBC World Service radio.

Each day we'd take an issue, story or talking point and invite contributions from listeners. They were a geographically diverse bunch. Some of our biggest audiences were in Nigeria, Sierra Leone, Liberia, Kenya, Uganda, Jamaica, the US and the UK. And the World Service's overall audience is, as you'd expect, sizeable. Given that, you might imagine that each time I'd start a programme, I'd ask a question and instantly be deluged with messages and calls. Not necessarily.

Our subjects would come from stories and issues from all over the world. Those stories would be hugely important but in some cases something that felt urgent to Ugandans and Kenyans might feel a lot less so to listeners in the UK or US. If I asked open questions such as 'Let us know what you think' or 'What would you like the government to do now?' or 'Are the protestors right to demand the President stands down?', the scale of the reaction might not be huge.

I then started to notice a pattern. I began directing my questions more.

If we go back to our imaginary story involving Uganda and Kenya, if I asked 'Whose fault is this?', the response might be muted. But if I prefaced it by saying 'If you're listening in Uganda and Kenya' and asked the same question, I'd often see a bigger reaction – even though by saying so, I'd ruled out most of the audience who weren't in either country.

I continued to experiment. Often, we'd be talking about global issues like climate change. Again, instead of asking open questions, I'd say, 'If you're listening in the US, what do you want your government to do?', 'If you're listening in Australia, how are you affected by rising temperatures?' and so on. Again, the more I directed the questions, the more people felt I was talking to them and the more engagement increased.

If you listen to great phone-in hosts like Shelagh Fogarty, Iain Dale or Stephen Nolan, they do this a lot. They talk directly to the people they want to hear from. They make it feel personal.

Once you start putting your mind to tailoring your message to your audience, you'll find you can do it in all sorts of situations. I am often asked to speak at conferences or to a particular news organisation. With these audiences, there is certain to be a huge range of people and professions. In that situation, I'd first try to establish who was in the room and then address some of the different people there. 'For those of you who make podcasts . . .', 'For those of you who work in long-form video . . .', 'For those of you doing live audio . . .', or whatever the case may be. Or if I'm speaking at a university and I know that most attending study history, I might say, 'I know you all study history – this next point connects directly to how you will approach sourcing information.' This is a much less rock 'n' roll version of when the lead singer of a band addresses a stadium crowd and says, 'Who's here from Manchester? Who's here from Liverpool?' and each section of the crowd cheers in turn. Or when the singer goes, 'Hello, Wembley!' It's designed to signal: 'I'm aware you're here, I'm glad you're here and I'm tailoring what I'm saying for you.'

My most potent example of how well this can work is the 50:50 Project, which, as I mentioned earlier, works to diversify contributors in media content. I decided from the start that I would commit to communicating in person and on email directly with as many participants as need be. That meant countless talks at small editorial meetings, one-on-one conversations with editors or producers, email exchanges on small details and so on. It was a lot of work, but it was vital to the project's levels of engagement and growth, especially at the start when it wasn't even certain I could make it into something

There is a flipside to this issue too. If you give people a chance to think 'This isn't for me', they may well settle for that conclusion. As such, when I'm communicating, I'm not just thinking of those I want to engage with but also those who are most likely to feel disengaged.

Here's an example from close to home. Throughout my children's time at their primary school I've popped in to do a 'news assembly' every once in a while. Most of the older half of the school pack in to one of the halls and the kids ask me anything they like about the news. It's a lot of fun and a vigorous test of how well I can really explain the news. Every time they catch me out. The children in the hall range from around seven years old up to eleven and, inevitably, the older kids are more confident asking questions. Once when I was finishing, I noticed some of the younger ones were getting fidgety. I would have got fidgety too.

The next time I went in, the questions still largely came from the older children but this time I made sure to calibrate my answers much more in the direction of the younger children. Two things happened: the younger ones were more engaged, the older ones remained just as engaged.

There is a direct connection between how well you communicate to an audience that this is for them and their level of engagement.

Now, by this point, we've tailored what we're saying, and we've made every effort to make it clear this information is for our audience, but there's still one possible catch. What if the audience isn't convinced by you?

5. Believing in the messenger: how best can you be credible?

If I'm trying to explain something to someone, I need to be credible to them on the given subject. If I'm not credible, they're going to be far less sure of what I have to say.

Stand-up comedy is a particularly brutal arena for credibility. If you're an established star playing in a large arena, you're already good to go. People have bought tickets to see you and they've done that because they think you're funny. When you walk on, the audience are primed to laugh already. If, however, you're a stand-up at an open mic night, the audience has no idea who you are and you have to earn their trust. Each time a joke lands well, you steadily become more credible. I can think of many times when a stand-up walks on, you can feel the nervous tension in the crowd, and then everyone relaxes once it's clear they're in safe hands.

In quite different ways this plays out in many situations where the person addressing an audience is not known – a trainer on a course, a new teacher with their class, a new coach for a football team, the new head of a department, a new physio treating an injury, a new columnist on a newspaper. Or if you're in marketing and are trying to position a new product, a lot of your effort will go into establishing both the company's and the product's credibility in the eyes of potential buyers. You'll be able to think of any number of examples where you've experienced this either as the person seeking credibility or the person deciding whether someone has it. Sometimes people's reputation and a company's reputation goes before them and that will help a lot. Sometimes they are starting from scratch. Either way, credibility is needed if they're going to have a chance.

Being conscious of what that credibility rests on and of what language you want to use to convey it gives you a far better chance of being heard.

I certainly found this with the 50:50 Project. I made sure that those of us making the case to journalists and producers were journalists and producers ourselves. This meant if someone said, 'It's particularly difficult for us,' we were in a position to say we understood because we faced those difficulties too.

Needless to say, we can't always be credible before we talk to people. We may be addressing people who know nothing about us. We may be addressing people who actively think we're talking about something about which they know more. In these situations, it's important to think of ways to establish your credibility.

EXAMPLE

This was an issue when I started to think about my explainer videos. They've been more successful than I ever dared hope, but back in 2019 they were an idea sketched out on three or four sides of paper and their success was far from assured. That I was able to get them going rests partly on how much I was able to build credibility in the eyes of the audience and of my colleagues.

I've gone back over a load of old emails looking for what I was saying about this idea. This is a section of one of the earliest ones I can find.

> What people want above everything, I think, is crystal clear explanation, high-protein information, sharp, pointed analysis of what has happened and a willingness to call out falsehoods. And to do so in a way that is authentic, authoritative and efficient. I think we can do this.

I finished off by arguing: 'It's often assumed you can't do this in a bulletin and be impartial. We can.'

This was a proposal to, in its own way, take a quite different approach to presentation and news analysis compared to the programming the BBC usually made. It involved the presenter doing both the reporting and the analysis. It also involved a more direct approach – 'assertive impartiality' in a phrase coined by my then colleague Gavin Allen – that brought with it additional reputational jeopardy. Not to mention, there was no guarantee the audience or my colleagues would like it.

Two years later, our videos had racked up millions of views and the *Sunday Times'* Rosamund Urwin had declared them a 'new genre of reporting'. This didn't happen by me having the idea, concluding I knew it all and ploughing on. On the contrary, in those two years I painstakingly built my credibility with the two audiences that mattered – my BBC colleagues and bosses and, most importantly, the viewers. It's perhaps useful to outline how.

I assessed the credibility that I was starting with

It's not particularly easy to assess one's own position, but in 2019 I was, I'd say, a mid-ranking BBC News presenter. If I was known for anything, it was probably for the fact that I used a large touchscreen. The challenge for me with the explainers was that people expect certain things from a BBC News TV presenter and I was about to do something different.

I needed to reassure colleagues and viewers that what I was doing wasn't undermining BBC News' commitment to be fair and impartial. And also, why should the viewer and colleagues see me as the person to do this? They probably wouldn't be expecting it.

To try to address this, I sketched out a plan.

I talked and listened to a lot of colleagues

I began by setting up conversations with people whose support I'd need – my immediate TV editor, the head of our department, the producers of my show and colleagues from the BBC News website. These weren't one-off conversations, they were ongoing. I approached all of them armed with as a clear an explanation of my idea as I could produce. And as I made my case, I also listened to them and adapted it. I was building the credibility of the idea, while also improving the idea through the input of my colleagues.

I developed an editorial process

Part of how I'd produce the video explainers would be that each script would be seen by several senior people and would go via an expert on that particular subject. After that, if there was any doubt, we wouldn't do it. We'd always err on the side of caution. But if it was good to go, then it was ready because of the help and support of several brilliant colleagues. Their credibility was part of the product's credibility.

After each video was released, I'd send it to a reasonably long list of senior managers, editors and journalists across the BBC (some of whom had not been involved in making them) and encourage them to critique what I was doing. This raised awareness of what I was doing and, most importantly, improved the product too. It also meant that even though I was doing something new, I wasn't detached from the main organisation.

At first, we'd only use the videos on my show and on my Twitter feed. Month by month, the number of BBC outlets who were interested increased. Now the videos are used across BBC output. Every week, that's a consequence of decisions taken by colleagues across the organisation on each new video. All of it rests on my credibility, the credibility of the brilliant team who make them and of the product itself.

I talked to the audience

I was in touch with the audience, too. Initially the videos were greeted with some surprise. When that surprise was registered on social media or in emails to the BBC, whether positive or negative, I'd engage and explain what we were doing. Video by video, what started as something different morphed into what was expected from me. And with each video that proved both popular and fair, my credibility to do this kind of explanatory analysis grew in the eyes of the audience.

I built credibility into each explanation

Everything I've detailed so far was the long-term work of building credibility with colleagues and viewers over months and years. However, I imagine you are thinking, 'OK, but what about when you can't build credibility over a long period of time?' Many of the circumstances in which we speak involve communicating with someone who knows nothing about you.

With my explainer videos, my answer to this was to build evidence of my credibility into the explanation itself. There is every chance someone watching one of them doesn't know me from the TV and has never seen one of these videos before. For this person, I am starting from scratch. To tackle this, I constructed the videos in a way that would build credibility there and then. If I was going to bluntly assert that something was correct or otherwise, I'd position supporting evidence right next to the assertion. This would give my statements an immediate factual root and help me build credibility in the moment. With the videos and other situations too, this would be my credibility checklist:

- **Am I credible to the people I'm addressing?**
  How do I want to be seen and am I achieving this?

- **Who do I need to be credible to?**
  Can they be treated as one or are there different groups?
  For each group, are there long-term ways of building my credibility?
  If yes, what form would that take? If no, how can I build my credibility quickly?

- Which aspects of my experience and knowledge will enhance my credibility?
  Can I speak on these areas with fluency and precision?

As I'm writing this, I am thinking of an article I read a few years ago. It was about how politicians increase their appeal to voters – but it leaned on some advice for job interviews: think about why they won't offer you the job and how you can address those doubts.

This ties into credibility. Whether it's a one-off, like a job interview, or something long-term, such as an idea or product or business you're trying to get going – ask yourself: where's the credibility gap? Then you can work up ways to address that.

Credibility matters. And every time we seek to explain ourselves, there are opportunities to use language, information and tone that can increase our credibility there and then.

KNOWING YOUR AUDIENCE: IN SHORT

There was a lot to take in there but hopefully it got you thinking about not just what you're saying but who you're saying it to. Here are those five questions again.

1. THE TARGET
Who am I talking to?

2. KNOWLEDGE ASSESSMENT
On this subject, what do they know and what would they like to know?

3. TAILOR IT
How do they like to receive information?

4. MAKE IT PERSONAL
How best can you convey that this information is for them?

5. BELIEVING IN THE MESSENGER
How best can you be credible?

THEORY INTO PRACTICE

You've now been on the receiving end of a decent amount of theory. I've always found my best work is done when I've thought long and hard about the reasons why I'm doing what I'm doing. Of course, I'm not expecting any of you to read through the whole first section of this book any time you need to explain something but I hope the aspirations and reasoning I've set out give you a good foundation as we move from theory into practice. Inevitably, some of what I've told you will resonate more than other parts. Hold on to those thoughts that really connect with you. You'll already have seen that I've collected all sorts of examples along the way that may have been inconsequential to others or had nothing directly to do with explanations – but for me they made sense in terms of how I can communicate better. You'll have your own examples that make sense to you. They might come reading this book, they might come as you think about explanation or at an unexpected moment. Whenever they do come, don't let them pass. Jot them down or make a mental note. They'll help shape and structure how you see information and how you want to pass it on.

Much of what I've described so far will, I hope, in time become habit. You have criteria that you can work towards when you're explaining something. You have certain considerations about who you're speaking to. Next, we're going to build some systems on top of this. These for me are essential if we're to turn our aspirations for high-quality explanation into something that is practical, adaptable and usable.

As we make this shift from theory to practice, you'll no doubt have a question or two.

'Simplicity', 'efficiency' and 'precision' would all appear to complement each other. 'No distractions' fits with this. As does making your explanation 'engaging' and 'useful', as well as maintaining 'clarity of purpose'.

How, though, do we do all this while also including the 'essential detail', 'complexity' and 'context'?

How do we include enough information to ensure a fair, helpful and informed explanation while excluding enough so that people stay engaged?

This conundrum, as Tim Bale, Professor of Politics at Queen Mary University of London, put it in a tweet, is: 'the balance of compression and comprehensiveness'.

Can we find that balance while giving due consideration to who this explanation is for and how they might like to receive it?

All this can feel like an awful lot to consider. It is! But there is a way through this that can take you from perhaps feeling daunted to explaining yourself brilliantly in whatever circumstances you are operating.

At this point, I feel like I should manage expectations. I am not about to do the book equivalent of a drum roll and then, ta-da, here's the answer. My experience over the last thirty years is that becoming better at explanation doesn't work like that. Instead, quality explanation is often the consequence of many small decisions working in sync to create a cumulative effect. When I am working on a particularly important report or speech or presentation, I frequently experience a nasty moment when it feels like no breakthrough is coming. I will have assembled a large amount of information, but I can't find coherence and clarity on what I should and shouldn't include or find the narrative to help me shape what I want to say. (I can assure you these precise sensations have been experienced repeatedly at various stages of writing this book.)

As I've got a little older and wiser, I've learned to navigate those moments by trying to put the worry to one side and, instead, to work through a system which, time and again, has helped me to get there in the end. I know if I keep doing the right things, I'm likely, if not guaranteed, to be able to explain myself well. Reaching that point is hugely rewarding. All the different steps and the small adjustments can, in isolation, feel like a lot of bother. I can think of times when someone has asked to see how I edit a video script or a news sequence for the TV and, as I show them, you can see in their eyes them thinking, 'Why is he obsessing over details this small?!' Not just factual details, but the smallest of details around which words I use and how I use them. I do this because the details add up to something that is more than the sum of its parts.

This detailed and thorough approach is very much what I would advise as you prepare for important moments – meetings, interviews, presentations, sales pitches, dissertations, essays, exams, speeches, briefing documents, lessons, staff meetings and so on. I call it 'Seven-Step Explanation' and you'll find it in detail in the coming pages. As I suggest what to do, I'll also take time to explain why I'm doing that. As I've mentioned before, I think you're much more likely to find this useful in the long-term if I both outline what to do and why I think it's worth it.

I am, however, mindful that once the system becomes familiar you won't want all that detail every time you look this up. So each section is summarised too.

There are also many types of explanation where you don't need to pore over every last detail of the Seven Steps. These more day-to-day moments still require us to be aware of what we want to say and how – but there are easier and more realistic ways to prepare for this. Often, for smaller explanatory tasks, the full monty is definitely not necessary. In that spirit, there's a guide to shorter verbal and written explanation towards the end of the book.

With all that said, let's get into it.

3

SEVEN-STEP
EXPLANATION

This is what the process looks like.

1. SET-UP
2. FIND THE INFORMATION
3. DISTIL THE INFORMATION
4. ORGANISE THE INFORMATION
5. LINK THE INFORMATION
6. TIGHTEN
7. DELIVERY

These words won't make too much sense on their own yet, but these are the Seven Steps we're going to work through. We'll take them one at a time.

STEP ONE: SET-UP

Let's assume you know you're going to have to explain yourself in some detail in a scenario that is coming down the track. For the moment, we'll assume it's a controlled scenario, such as an essay, a report, presentation, a speech, a lecture or a work briefing where you will have an opportunity to speak or write at some length.

(We will look later at unknown or dynamic scenarios such as conversations, interviews, meetings and so on. The work we're about to do will help with those too. But for now, we'll work towards something where we will have the chance to say exactly what we choose.)

We're going to begin with a series of questions, all of which can be answered in a sentence or two.

- **What do you hope to explain and/or communicate?**
  Provide one sentence maximum. If you're struggling, write a paragraph with everything you think is the purpose of this explanation. Read it over a couple of times. How would you summarise the overall purpose? Have another go at writing a single sentence. If you're still struggling, try writing a series of single sentences outlining what you think is the purpose. Does one feel more important? Perhaps it works as an overall purpose for what you're trying to do. If not, is there someone you can run it by? A colleague, friend or relative? If you can get your purpose for an explanation clear, it will be a huge help in any number of decisions you're about to take.

The next questions will be familiar from the previous part of the book.

- **Who is this explanation for?**
  Give the answer in one sentence maximum.

- **Is there a consistency of knowledge amongst those you're addressing?**
  Yes / No

- **How do you assess their knowledge of this subject?**

- **How would you summarise what they'd like to learn from you?**

- **What specific questions will this explanation need to answer?**

- **What, if anything, do we know about how they like to receive information?**

  If possible, provide a brief number of bullet points for each of the last four questions above. Are you confident answering them? If yes, fantastic. If not, you're light on information about who'll be receiving this explanation, so you'll need to think about the next question.

- **Are there ways you could find out more?**
  Provide a few bullet points.

- **Where will this be consumed?**
  One sentence maximum.

- **Is there a fixed duration?**
  Specify a number of words or minutes.

- **Is the duration strict?**
  Yes / No

EXAMPLE

This is how I might approach a speech at a journalism college.

Who is this explanation for?
— A group of around thirty journalism undergraduates.

Is there a consistency of knowledge amongst those you're addressing?
— Yes

How do you assess their knowledge of this subject?
— Strong interest in journalism and the media but still learning about the practicalities and principles.
— Will be keen consumers of the news so will have awareness of main stories but not necessarily detail on all of them.

How would you summarise what they'd like to learn from you?
— Keen to get insight into the practicalities of producing journalism and working in the media.

What specific questions will this explanation need to answer?
— How I became a journalist.
— How we make one of our video explainers.
— How I avoid bringing my opinions into my journalism.
(This is just the start of this list – we'll expand it later.)

What, if anything, do we know about how they like to receive information?
— Not completely sure, but they will want to learn a lot rather than just enjoy the talk. Need to make it as practical as possible – with as many takeaways as I can offer. Need to clearly signpost the most important

lessons to make it easy to spot them and understand them.

Can you answer the last four questions?

— Not completely.

Are there ways you could find out more?

— I'm going to email some questions to the lecturer who organised the talk so that I better understand the course they're on, the specialisms they're studying and any particular areas they specifically want to hear about.

Where will this be consumed?

— An in-person talk in a lecture theatre at the university.

Is there a fixed duration?

— Forty-five minutes.

Is the duration strict?

— Overall yes. The split between my talk and the Q&A is loose.

Most of the time, answering these questions will take less than ten minutes. You might be thinking as you read them, most of these are so obvious, why bother? You'd be right to say they're obvious. There are no questions there which are outside of basic considerations. But – and it's a really big but – human beings have a long track record of not stopping to ask themselves obvious questions on what they're about to do. Frequently, I'll be going through these questions, and I'll think, 'Oh, of course I should include that' or 'Of course I don't want to do that.' It's obvious when I think of it. The problem is that I hadn't done!

The idea of the 'Set-up' is to make us conscious of what we're about to do. It helps us set the goal we're working to. If you go through these initial questions and all the answers were already in the front of your mind, then no harm done. More often than not, though, it'll help spark ideas and clarity on what you're about to do. Explanations don't work in a vacuum; they work in specific circumstances and to specific goals. The 'Set-up' makes sure we're clear on what those are.

> ### QUICK CHECK
> Are you happy with what you're doing and who it is for?

If you are, it's time to start collecting information for our explanation.

STEP TWO:
FIND THE INFORMATION

Starting the process of constructing a detailed explanation can be daunting, whether you are knowledgeable on the subject or not. I take on explanations most days at work and it can still feel a little overwhelming at first.

Where do I begin? Can I really become sufficiently expert in this to be fluent? Can I really get *all* of this across in a way that makes sense? Maybe this one is too hard to explain? Is there time to get through everything? It can feel like standing at the bottom of the mountain looking up.

I used to get those feelings every week when faced with my latest history essay at university. I get those feelings every week now whether it's because of a video explainer or a speech or an important meeting.

In this situation, I try to take a deep breath and simply start. And by start I don't mean trying to crack exactly what I'm going to say. You can defer the problem of having to be concise and coherent on the subject! Instead, let's just begin gathering information that may be relevant. The extra level of knowledge – of knowing whether information *is* relevant – can prove elusive and it's completely normal not to be sure at this stage.

It's worth adding that, unlike, say, an essay or a presentation on an unfamiliar subject, many times we have to provide an explanation on a subject that we know well. Indeed, it might be what you spend every day working on. You might be making a pitch for a product you helped to develop. Or perhaps you're a drama or music teacher who's about to introduce a new piece or play. In those situations, you are going to have a good handle on the available and relevant information. But assuming this is a new explanation you're working on, let's continue nonetheless. Our task here is not how to order the information, but simply to put everything we think may be relevant into one place.

(You might like to know that work on this book started in a Word document called 'Info Dump'. Whatever the merits of what I'm telling you, I do practise what I preach.)

How much information you collect will in part be decided by the breadth of the subject and how much information is available. If you've the time, err on the side of too much. You can always delete it.

Start by asking yourself:

- **Where and how should I look for information?**
  Before you launch into gathering information, it's always helpful to have a rough direction of travel.
  You know your 'purpose', but you may not know the subject particularly well. Nonetheless, start with this question.

- **Which parts of the subject do I want to explain?**
  Make a list of bullet points. Don't worry if you include subject areas that may turn out not to be relevant.
  Or if you miss subject areas. That's understandable if you don't know the subject well. Just put down what you think are the areas where you'd like to start.

And then the hunt begins.

A FEW WORDS ABOUT SOURCES

Where you look for information will, of course, be shaped by the subject that you're seeking to explain. But there are some guiding principles that we can keep in mind as we assemble the information that we need.

First, we need to build a factual foundation on which we can establish our explanation. These facts are the core of the subject. To find these facts, I would normally turn to what we can call 'direct sources' and 'reliable sources'. A 'direct source' meaning we have the information direct from the people who'd know best. If you were explaining the weather, it might be the Met Office; if we have a doctor's appointment, it's the test results from the nurse; if we're talking about a company, we have its publicly filed accounts. There may be dispute over what to make of this information, but in as much as the information is very likely to be accurate, direct sources are helpful.

Then there are 'reliable sources'. These are sources of information that are not the original – or primary – source of the information but are intermediaries who you have very good reason to trust. If you run a small business and you want to explain how recent government announcements may affect your business, you might refer to an industry body or to a media organisation whose assessments are routinely reliable. Given their track record of being accurate, it's reasonable to assume that the information is just that.

When I'm starting to assemble information I always start with direct or reliable sources. I also keep track of where I've taken the information from. You want to know where each fact has come from should you, down the track, need to check something about it. Also, depending on how you use the information, you may want to source it within your explanation. For example, in a slide in a presentation, if you pull up a statistic, you'd want to note the source of the statistic. If you're writing an essay, you will need to add a reference for the quote you've used.

You may also want to double-check the context in which a certain piece of information appears. One example of this would be when we're looking for quotes for our explainer videos. Often, we only want quotes that are one or two sentences long. So before we use them it's wise to double-check the context in which those sentences were said. Having a link to where they originally came from allows us to do that easily.

There will be some facts that need no further checking. Perhaps it's the date on which a company appointed a CEO. If the company provides that date, we can work on the basis that it is correct. There will be others you want to be more certain of. A great rule of thumb is the BBC World Service's 'two source rule'. This means that you want the information from two reliable sources. Now, what is deemed reliable is clearly a matter of judgement but again I'd take my lead from the BBC World Service.

My colleagues there will judge information sources on their track record. If it's a source you've used many times and it's always been accurate, then you can reasonably conclude that the information is likely to be true. And if you have two reliable sources saying the same thing, that is a good basis on which to proceed. Again, though, keep track of the sourcing of your facts.

If you've pulled together the factual foundation of your explanation, then you'll also be wanting to bring extra texture and perspectives to whatever it is that you're explaining. This may involve using information sources you know less about or where there are known reasons that the information may be slanted towards a particular position. This doesn't mean the information has no value, it just means you need to handle it with particular care. It's always worth asking:

- Have others raised concerns with this source of information?

- Is there an agenda or belief that I need to factor into my considerations of this information?

- What do I know about the organisation or platform where the information is appearing?

- Where do they get their information from?

- Can you see people you trust using this information?

- Is the information being shared by this source also being shared by other sources? Does what they're saying match?

All these questions will help flush out risks that come with the information.

And I've three more questions to offer you which I use all the time at work and in my day-to-day life.

What's the source?

I'm not sure how impressed our daughters are about it, but this question has become a catchphrase in our house. My wife, Sara, and I started using it when the girls would come home with a claim about something that was going to happen at school. Often when we asked where the information came from, it was from a friend. When we asked where *they* had got it from, no one was really sure! Saying 'What's the source?' became a very effective way of quickly working out the quality of the information. If no source can be found, that is clearly a reason to raise an eyebrow. If we can find the source, then we're into assessing how reliable it is. Just like news sources, whether a piece of information had come from one friend or another would impact on how we assessed this. One may have a track record for being reliable, another less so. We make some of these calculations subconsciously, but the more conscious we make them, the lower the chance we'll rely on or use duff information.

Once you start doing this in day-to-day life, you will notice how much information is being shared where the sourcing is either non-existent or unreliable. In an era of 'fake news' and wild online rumours, 'What's the source?' is an essential phrase.

As you assemble information for any explanation, always keep this question in mind. If any information source is stating something with certainty but does not show you where that 'fact' comes from, immediately be sceptical. The 'fact' may be true, but you want to do more to find out if it is.

Really? (asked with a raised eyebrow . . .)

The second question is a one-worder and one I've really come to rely on. If some of the information you're gathering doesn't add up to you, don't ignore that. For two reasons. First, if it doesn't add up, it might be wrong. Second, if it doesn't add up to you, it might not add up to someone you're sharing it with. Indeed, you may want to acknowledge that when using it – 'and, though you may find this hard to believe, it's actually the case that . . .'

It can be easy, especially when you don't know a subject well, to ignore your brain saying, 'Really?' When I'm collecting information for an explanation, I'm asking it all the time – as I do when I'm being given information in my day-to-day life. Trusting your instincts is a very good way of catching bad information.

Who else would know?

If I've already asked 'What's the source?' and 'Really?' and I'm still not sure of the information I have, then I get help. I think about who could guide me on whether the information I'm concerned about is correct and can be trusted. Who you may be able to turn to will, needless to say, be dictated by what you're trying to explain but most of the time there will be someone. If you're writing an A-level essay, your teacher would be the first port of call; if you're talking about one aspect of your industry, you'll know other specialists; and so on. I know it's far from guaranteed that you will have access to the perfect person to help, but it's always worth thinking about who you could turn to. You can be sure you won't be the first person to have considered the merits of the information that you're wrestling with.

With all these considerations in mind, as you look far and wide, deposit anything you find that may be relevant into your document. If in doubt, put it in. If you are copying and pasting, make sure you label the source and author at the top of the text and put a link in too.

YOUR QUESTIONS LIST

As you pull together your information, as well as asking questions about the trustworthiness of what you have, also have in mind the questions you and others may have about the subject.

What do you think people won't understand?

I always try to second-guess the parts of my explanations where people might get lost or confused. This might be because it's a particularly complex subject or it might be that what I'm describing doesn't, at first glance, seem logical. If you're alive to these risks, you're much more likely to address them.

Use your questions in the 'Set-up' as a starting point, but as you collect information keep adding questions. You may have one, you may have twenty. What matters is that you're sensitive to the questions people have about the subject generally and to the moments where an explanation may become confusing or appear contradictory. It's hard for me to overstate the importance of this list. When I'm towards the end of work on an explanation, one of the most important moments is to go back through this list and check, one by one, that every question I can think of has been answered. The more people have their questions answered, the more interested they are going to be and the more value and credibility they see in you as a source of information.

What don't you understand?

I love this one and use it all the time. Kidding yourself that you understand something when really you don't is likely to lead to trouble later on. As discussed earlier, we need understanding to get to the best explanations so, I'm afraid, there's no use in trying to short-circuit the process. My BBC colleague Mary Fuller is brilliant at this. 'I just want to spend a bit more time on this,' Mary will say. She'll have spotted a section of one of our videos which is lacking clarity and conviction. It'll almost always be because we haven't really understood it. Mary will say, 'Give me twenty minutes' and will focus directly on this one issue. What we end up with isn't longer but it's better. Sometimes this will happen when you're a long way into an explanation but, where possible, I try to meet trouble halfway. I've long enthusiastically embraced my lack of understanding and am on the lookout for it.

If I'm having to explain a subject on which I am far from expert, this is an essential list. But even with subjects that I know well, there are always areas where I could be sharper.

I start this list at the same time as I start my document where I'll be putting all the information. As I explore the subject, each time I feel myself getting confused I write down the issue. I might try to find an answer there and then and put that information with the rest. But it's possible I have the right information on the subject but just don't understand it. I don't fixate on that. My goal here is to collect the information and note the areas where I'm confused. Alleviating that confusion will come in time.

Keep this list. You're likely to add to it further and we're going to use it all the way through the Seven Steps.

Have you discovered subject areas where you'd like more information?

We want to make sure we're covering all bases. If you're not sure you have what you need, turn back and look for more. Better to quickly go and get it and discard later, than not to consider it.

- Add any areas of missing information to your original list.

- Go and find information on those subjects and add it to the pile.

- Go down through the whole list. Do you have information on each point?

- If yes, fantastic. If no, what's the plan?

If you're struggling to find information, remember that there are almost always people who we can ask for guidance. It's rare that people mind being asked for their expertise if it's done in the right way.

A simple example of this has been how we've supported our daughter Alice through her exams at school. Several times she's come to me with variations on, 'I'm not sure if I need to revise this – I don't know if we're supposed to focus on this textbook or this handout.' I guess if I put the time in, I would eventually fashion an answer to that. A much better and quicker option was for Alice to ask the teacher of that subject. This might seem obvious but a lot of children, Alice included sometimes, might not spot where they need help or how to ask for it. So, as Alice was revising, I suggested she had a piece of paper where she wrote questions that she had about the materials she was learning. On some subjects there would be none, on others there would be several. And on those subjects where she wasn't clear, she'd go to find the teacher and ask them. It's a simple thing – like many ideas in this book – but when we systematically ask for help every time we spot that we don't understand something, it can be very powerful.

OK, let's pause to consider where we've got to. If our set-up is the foundation, the information we gather is the first floor of our construction.

I'm not going to put an example of how this collection of information might look as I'm not sure it would be much fun to read. Suffice to say, if I did, it would be disorganised, incoherent and – if it was for a detailed explanation – long. That's completely normal.

QUICK CHECK

By this stage we should have the following:

- A summary of what you're trying to explain and for whom
- A list of the questions you think the audience will want answered
- A list of what you don't understand – or want to better understand
- A list of the subject areas you think you need to hit
- A pile of information on the subject

All being well, you now have a clear brief and your raw materials. Don't fret if that pile of information looks large and unwieldy – we're about to deal with it. On to Step Three.

STEP THREE:
DISTIL THE INFORMATION

I'm from Cornwall in the south-west of England. In the eighteenth and nineteenth centuries, Cornish mining was a big deal. For a time, it produced the majority of tin in the world and Cornish miners travelled far and wide to share their expertise. Nowadays the mines are closed but there are some you can still visit.

Geevor Tin Mine sits atop the cliffs of the Cornish north coast, just before the north and south meet at Land's End.

The section where the ore was processed is called 'The Mill'. Here the ore would have been refined from huge chunks that were pulled up from the ground to a much-reduced distillation that eventually became an ingot of tin.

The Mill is a magical place. It sits silent now, bar the sound of the visitors and the guides, but it's not hard to imagine the constant movement and noise when it was in full swing. All around are machines – including the fabulous shaking tables that lived up to their names. It was the work of these machines to turn the earth's raw materials into a valuable resource. Everything in the mill is coloured terracotta from the years' worth of dust produced in pursuit of that goal.

The main building contains a machine called a ball mill. It's the first stage in the process of refining ore: a large drum with big steel balls inside. The ore is added, it spins around, and the ore starts to break down. Without the ball mill, all the other techniques to distil the ore that follow couldn't be done. You can probably see where I'm going here. We need to go through the same process with information. If you want to organise and explain information effectively, you're unlikely to be able to do it in one go, especially if there's a lot of it. Certainly, I can't do it in one go. Instead, I use different processes for the different stages.

We're at the ball mill stage now.

The purpose of this distillation is to break our information down to its absolute minimum – the smallest nuggets. We want each element to be in its clearest and most usable form. To do this we need to strip out anything that is not essential to understanding these elements. While doing this, you don't need to worry at all about ordering the information. We're trying to do two things here: refine information that is or may be relevant and discard information that we definitely don't need.

THE FIRST SWEEP

Here we go:

- **Remind yourself of your 'purpose'**
  We need to keep this in mind at all times. This isn't a test of whether information is interesting – it's a test of whether it's *relevant and essential.*

- **Start assessing your information**
  No need to overthink this. Just begin with whatever you have at the top.

- **As you read, ask yourself: is this relevant?**
  Line by line, paragraph by paragraph, keep asking this as you read.
  If the answer is no, 'cut and paste' it to either the bottom of the document or a new document. (I'd recommend the former if it's a relatively short explanation; if it's anything longer, a second document is easier.)
  If the answer is yes or maybe, keep it.

- **With each section that you decide to keep, ask yourself: what is it that is of value here? And then start to remove everything else.**
  Focus on the information that has value and strip everything else out. You don't need sentences around it. (Indeed, you don't need *any* full sentences unless they are part of a quote or a passage that you explicitly want to keep in its entirety.) Don't skip this part of the processing. If you leave a full paragraph which contains a single point or fact, you will only have to go back to it later on and undertake the same process. Much better to do it here. The only time I would leave something alone is if I'm genuinely not sure if I need it. I'm hoping I'll be more confident to make that judgement soon – in the meantime, I'll leave it in its original form.

EXAMPLE

Here is a paragraph on the Battle of the Somme from BBC Bitesize.[7]

> The Battle started on 1 July 1916 and on that day the British army suffered its largest number of casualties ever – 19,200 dead and around 60,000 wounded or missing. Most of the casualties fell in the first hundred metres of no man's land.

Here is what I'd want to be left with.

> 1 July 1916
> British army largest number of casualties ever
> 19,200 dead
> Around 60,000 wounded or missing
> Most fell in first 100m

As you work your way through the information, a long list of facts, phrases, arguments, quotes, statistics, graphics, theories and so on will start to emerge. These elements are whatever you think has value for your explanation, in their simplest form.

As you go through your research and begin distilling it, all being well something pleasing will start to happen. Slowly but surely, you'll begin to get a better feel for both the subject and what you want to say about it. Slowly but surely, it will start to feel more familiar and you'll start to see patterns in how the information connects. And you'll reduce the size of your information pile by well over half, maybe much more.

STEP THREE

THE SECOND SWEEP

Now we're going to go back to the top and start again. This is far quicker than the first sweep because you've already discarded a lot of information and most of what's left has been reduced down. On this second sweep, you'll be in a better position to judge what you want to keep or lose. If you decide something no longer supports your purpose, delete it. Be strict on this – if something does not further your overall aim, it has to go. If you were giving an after-dinner speech or tribute to a colleague who's leaving, of course diversion and tangents are entertaining, interesting and worth doing. But if clarity of explanation is our goal, let's be strict about what we do and don't include.

HOW TO DECIDE RELEVANCE

There's no easy route to deciding what is relevant nor will this ever be a question with a definitive answer. Relevance is a judgement and it's quite possible two of us would draw different conclusions. But there are questions we can ask that will draw us towards conclusions that serve our explanation.

- **Does this particular element help me meet the 'purpose' of this explanation?**
  Sometimes, this will be clear-cut. Other times, you won't be sure. If it's the latter, move on to the next question.

- **Why does this particular element of the explanation matter?**
  If you're filling in a job application, if your overall purpose is to show you are excellent for the role, how specifically is this element helping you make the case? Is it demonstrating skills, insight or experience that match what the job description demands? Does this help to build evidence that you are a candidate who explicitly meets the requirements? Do these elements help make that connection?

  Or if you're writing a presentation making the case for change in your department, does the information offer explicit evidence to back up your analysis or to establish the context that makes the case for change more persuasive?

Every piece of information you include needs to be doing a job. If it isn't doing one, it's a distraction.

Again, if you're not sure, ask. In the years after the UK voted for Brexit, we have made any number of explainer videos on the ramifications of leaving the EU. I'd be at pains to make sure that I included the right information and frequently would not be sure. My then colleague Kevin Connolly, a brilliant foreign correspondent, was based in Brussels at the time. Often as I approached this point in the sifting of information, I'd email

Kevin. The subject would often read 'a favour' and Kevin knew all too well what the favour would be. 'Me again,' I'd write. 'I'm not sure if I need to include this statement from the European Commission . . .' or whatever the case was that time. Kevin would help me judge if my explanation was unbalanced or lacking context if I didn't include it.

If you can't get advice and are still not sure, leave it in for now. The priority here is discarding information that we're sure cannot help us.

As you read this, you may think that it's obvious to avoid information that's not directly connected to what we're trying to do. But, believe me, we all frequently devote time and words to things that don't really matter in the context of what we're trying to explain. You'll all have experienced times when you've answered a question in an interview or meeting or conversation and had that feeling of coming away from your moorings. Of drifting aimlessly and without any real connection to what you've been asked. We've all been there, and it doesn't feel good. You will have been on the receiving end too, when you're listening to a speech or some training at work and there's a sudden strong feeling of 'What is this about? And what has it got to do with me?' It's not that what you're hearing is in and of itself bad; it just feels irrelevant.

The better you get at sifting essential information away from interesting but non-essential information, the clearer your explanations and communications will be.

As you sift, your document will get shorter and shorter.

If I'm being sent on a reporting trip for a big set-piece story like a summit or an election, I often do Step Three on the plane. I'll open up the document, look at the page count, which is normally terrifyingly high, and wade in. The amount of information appears unmanageable but within two sweeps I am starting to take control of it and the page count has plummeted. I'm still a long way from being on top of things but I'm starting to feel more confident.

QUICK CHECK

- Are there any gaps in the information you need?
- If there are, repeat Steps Two and Three for where you see a gap.
- Do you have anything to add to the list of questions you have?
- Is all the information you have in its simplest form?

By the end of Step Three, we're in a strong position.

- We know what we're trying to explain.
- We know who we're explaining it to.
- We've completed a comprehensive sweep of information that is or may be relevant.
- And we've distilled that information, which makes it more usable for us and more consumable for our audience.

But these nuggets of information are only going to make sense if we create a structure for them. Explanations don't just offer information; they offer a structure to give that information meaning and relevance. That's the job of Step Four.

STEP FOUR: ORGANISE THE INFORMATION

By this stage, we're starting to see the subject more clearly. This is going to help us as we order the information that we've prepared. In Step Four, we need to identify what I call the 'strands' of an explanation.

'Strand' is the word that I find works best to describe the primary sections within my explanations. I use it to capture the idea that you are dividing your information according to different aspects of the subject. For me, just as there are strands to a story or an argument, there are strands within explanations. But, if you prefer, you could call them sections, chunks or themes.

Some strands will jump out at you as you consider the subject you're taking on. Others you may think you need but you're not sure. At this stage, let's include everything we can think of.

MAKE A LIST OF THE MAIN STRANDS OF THE SUBJECT

There's no fixed number of strands. Make a list of the ones that come to mind and give each a short header. The order of the strands doesn't matter and, while you want to choose carefully, the list can change as you continue your work.

EXAMPLE

This is a simple example of how I use strands. I had an important appointment with a doctor while writing this book. It was the culmination of six months of tests and its outcome mattered a great deal. I knew I needed to plan in advance for what I wanted to say, not least because I find in these moments of pressure having a plan helps me say what I need to, even if my mind is racing a little.

The appointment was important but not overly complicated. I ordered the questions and information I wanted to get across into three strands: my symptoms; my medication; and what activity I could and couldn't do in the future. Three was enough. In the few minutes I had with the consultant, this made sure I explained myself. (And when it came, the news from the doctor was good, which was a relief . . .)

That was an important but relatively straightforward piece of explanation.

I use the same process of dividing the information I have into strands if the explanation is much more substantial. For a ten-minute presentation at work, I might have five strands. For a thirty-minute speech or a long article or essay, I might have between five and ten. Whatever the scale of what you're about to do, selecting strands is an important step towards giving shape to that information.

Core strands plus two more

You should now have a list of the strands that you've identified. To these, we're going to add two more. One is for information that we're unsure how to use. One is for high-impact information that could be useful at the start and finish of our explanation. In your document you'll have something like this:

STRAND A

STRAND B

STRAND C

STRAND D

STRAND E

STRAND FOR HIGH IMPACT

STRAND FOR INFORMATION WE'RE NOT SURE OF

Soon we're going to want to start putting the information we've distilled into the strands – but not quite yet! Before that, we need to think about narrative.

WHAT STORY DO YOU WANT TO TELL?

As history teaches us, narrative can be used for the worst and the best of reasons. What is beyond doubt is the potency of stories. Tell a good one and people will hang on your every word. If you're a brand, shape a narrative (and even better an origin story) about your business or your products and it can be incredibly powerful in making people feel favourably towards you.

Professor Pragya Agarwal is a behaviour and data scientist. In an article in *Forbes*, Prof Agarwal has written of the importance of stories to businesses:

Successful brands create a narrative, something that positions them as unique, and creates an emotional connection with the customers. Good storytelling is all about forging an emotional connection.

The colours, the logo, the name and the tagline all follow from the central message. They all come together to form the brand story.

A brand story tells the motivation for starting your business, why you get up and do what you do every day, why customers should care, and why they should trust you.[8]

Here again we hear not just the importance of stories but also how their purpose – the 'central message' – guides everything else. It is the magic ingredient that gives everything you have to offer – information, design, products, customer service – a coherence and connection. It adds emotion too and, if it's a story people want to hear the end of – or even want to be part of – then your chances of people engaging with what you have to say spiral upwards.

This isn't only true of how businesses communicate with their customers. It's true of many types of communication in our lives. I understand when you go to the vet, you may just want to pass on the fact that your cat is limping. But there are many more examples – from teaching to media content creation to leading a sports team to sales – where using stories to deliver information is an essential tool in your armoury.

If, from the start, your audience wants to know what happens next, there's a very good chance that they'll stay with you.

I mentioned earlier, I'd called a then colleague Jonathan Marcus for advice on NATO. For many years, Jonathan was the BBC's defence and diplomatic correspondent. And in one BBC training course in the archive, he offers this uncompromising advice:

> Before you start to write and when you think you've finished, take a few seconds to ask yourself: 'What's this story in five words?' Have you conveyed that?

Jonathan's a hard taskmaster. I always felt five words might be a little too demanding, but I very much share the sentiment. We've already looked at the importance of clarity of purpose.

That is to understand what we hope to achieve by the end of the explanation. Clarity of story is equally important. This is about how we're going to deliver on the purpose. What is the journey you want to take your audience on? Where will you start? What will you promise them? We know the desired destination – but what route will you take to it?

The importance of narrative is no secret. Where perhaps we go astray is not understanding the centrality of narrative to explanation and communication. Arguably it is the most powerful tool at our disposal. Because while the information may be valuable, it may not be enough to draw people's attention on its own.

Pure facts and context are, of course, a wonderful resource but, for whatever reason, they don't grab our attention as a story does. They do not have us wondering 'What comes next?' Narrative does that. The best explanations deliver an abundance of relevant useful information via the best-told stories.

This is another lesson I learned through my own mistakes. When we started our explainer videos, we put what are called 'slates' at the start of each strand of the story. If we were making an explainer on Northern Ireland and Brexit, the strands might be 'Trade with the EU', 'The Good Friday Agreement' and 'Party Politics in Northern Ireland'. At the beginning of each section, you'd see those headers across the screen. We thought this would help guide the viewer through the subject. And, in a way, it did. But I started getting notes from people saying they didn't like the slates. One, the US media executive Vivian Schiller, suggested: 'I'd get rid of them. They're getting in the way.' Not for the first time, Vivian was right.

This was another of those moments where something clicks. Now, while I still look for ways to share the structure of my explanation, I make sure the telling of a story is the most important aspect of the videos. The videos got better as a result. One friend said to me the other day, 'It was seven minutes, but it was a quick seven minutes.' That's the story doing its work – inviting the viewer to wonder 'What happens next?'

Everything about how I assemble information for an explanation is guided by the story I want to tell. This has no negative impact on everything else we've discussed – relevance, precision, efficiency, structure and so on. It just means that you give narrative its best chance of supporting your explanation.

This is why at this stage in Step Four – where we organise our information – we must pause to think about the story we want to tell. Here are some exercises I use to help me settle on what that might be:

If someone called you now and asked, 'What are working on?', what would you tell them?

Humans are good at telling stories. If you're in the pub and someone asks how a recent work trip was, you're going to be able to answer fluently. It's the same for thousands of questions we ask each other every day. Our instincts aren't always right, but it's definitely worth checking in on them.

If you're struggling how to tell a particular story, imagine you had to tell it to a friend. What would you do?

Test out different starting points

If you have, say, five core strands marked out for your explanation, see if you can verbally outline how you'd talk through it all – starting with a different strand each time. Very quickly you'll feel that most of them don't work as a starting point. One or two might do, though.

Consider different story structures

There are many ways of telling a story. If you're stuck, you could consider these options, which are classic approaches to explanation.

Chronological: Use the passage of time to provide your structure. Break the story down into several sections that represent the key developments as they played out.

Finish / start / finish: This is a variation on chronological. You first outline the outcome you want to focus on, and then return to the start to work through to how it happened – until you're back at the outcome again.

Zoom out: Start on the event or issue you're focused on. Zoom out one step at a time to reveal more and more of the context and detail of the issue.

All context: You're explaining an issue or event. You establish what it is – and then say, 'but you can't understand this without understanding X'. Introduce a strand of the subject. Conclude it – and repeat. 'But even if we consider X, that alone doesn't explain this – because you can't understand this without understanding Y.' Bit by bit you build the context around whatever it is you're explaining. If 'zoom out' wraps a wider piece of context around the explanation each time, 'all context' takes a thematic approach giving equal scale to several pieces of context.

What someone said: There may be a statement or finding around which you can build your explanation. Could you begin your explanation with this, establishing a fact or turn of phrase – before unpacking each strand with that phrase or fact as a point of reference? You can return repeatedly to the wording for your structure and language.

Solving a problem: Establish a problem that needed solving and then section by section outline how it was addressed.

Block by block: Along with 'chronological' and 'solving a problem', this is the one I turn to the most. If you can create a sense of constructing an explanation in front of your audience, you can create momentum and also curiosity about where you're going. The idea being that each part of the explanation would not make sense without the part that has gone before.

These are some ideas – you may have others. As always, keep an eye out for what people do. The best explanations come with a story technique attached. If you can feel someone explaining something clearly to you, ask yourself what story are they telling you and how?

Try writing the first sentence

If all else fails, sometimes it's worth beginning an explanation – an essay, a talk, a seminar, a report – by stating explicitly what the story you're about to tell is. For example, in 2021, I made a long video explainer on the heatwave in North America. This is how the video starts:

> *This is a story of two heatwaves – that have set record temperatures. That have started wildfires. That have killed people. And that connect to what we're doing to our planet. And how we're tackling climate change.*

That, in fifteen seconds, is what I then spend the next ten minutes unpacking. I have laid out how I'll take on both the story of the fires right now and how they fit into the longer-term story of climate change. Through those ten minutes that is precisely what I do – section by section, expanding the focus from the immediate impact of the heatwave on places in Canada and the US to what is happening to our planet. It's an example of the 'zoom out' approach. Needless to say, different circumstances and different stories require different language. The way I started that video, though, is a useful device for you to use, especially if you're struggling to get going. How would you finish these sentences?

- This is a story of . . .
- This lecture will explain how . . .
- Today, I'm going to take you through . . .
- Here are ten minutes on . . .

Having this clear in your mind and written down is useful. There may be overlap with your purpose and that's fine.

If we take the heatwave example, the **purpose** is to explain what is happening with the heatwaves and how they connect to climate change. The **story** is how one Canadian town's plight reveals broader truths about what is happening to the climate.

Both the purpose and the outline of the story can be your guide. In both cases, if what you're putting into your explanation isn't working towards those twin goals, then you're off track.

EXAMPLE

Let's imagine I am a senior manager at a hospital which is bringing in a change to the IT system.

What is the purpose of your explanation?
To explain what changes are being made, when they're happening and why they're happening.

What is the story you want to tell?
That our healthcare is being impacted by out-of-date IT and that this change will make our working lives easier and improve the care we offer too.

How do you want to tell that story?
By contrasting the IT problems we have now with the successes in other hospitals that have already introduced this new system. I'm seeking to demonstrate those successes as I walk through the stages of the planned change of system.

ORGANISING THE STRANDS

Having set out our strands and thoughts on how we want to tell this story, it's time to order our strands. This may change in time, but let's have a first go at it.

They might look like this:

STRAND C

STRAND A

STRAND D

STRAND E

STRAND B

STRAND FOR HIGH IMPACT

STRAND FOR INFORMATION WE'RE NOT SURE OF

STEP FOUR

ADD THE INFORMATION

This is a really satisfying moment and it's when you will, I hope, feel your explanation start to take shape. Place your strand headings in one document and open up the one which already has your distilled information in it. (You can do it all in one document if you prefer but I find I am scrolling endlessly up and down if I do that.)

Start going through your distilled information and move each element into the strand of the story where you feel it fits. Given your choice of strands has been informed by the distillation process, most of what you have will find a home. If you really don't think something fits but that it is important, put it in the strand for just that. Also keep an eye out for elements that are so clear, relevant and high impact you might want to start or finish with them. At this point, it doesn't matter how you order the elements within each strand.

This process is such an important part of developing your explanation. It's a chance to really start testing how well you're understanding the subject and, just as importantly, how all the information you've collected fits together. As we assess the information and place it into strands, it's becoming more and more familiar to us. Continue this until all the information is in one of the strands.

ORGANISE THE INFORMATION IN THE STRANDS

Next, we're going to look at each subject strand in turn. Read through all the elements you have in it and ask yourself these questions:

- What do I hope each strand of the explanation will achieve?

- Within each strand, which of the elements are the most important?

- Which should I start with?

- Which elements do you think will follow on from another?

- Here and now, if you had to describe each strand to someone, how would you do it?

This last question is *particularly* important as our end destination here is a natural, engaging explanation. This is the start of the transition from a collection of information to a story and an explanation that people want to hear or read.

Having mused on all these points for however long you choose, start to order the elements within each strand. Start from the top. Place the element you want to begin with. Next, bring in the element you think should follow. Either in your head or out loud, outline how you would flow from one element to the next.

Step by step, build up an order of elements that meets your ambition for this strand of your story. With each element you include, make sure you are clear on the role it is playing in your explanation. Keep going until you feel this strand of the explanation has sufficient information to do its job. There is no obligation to use all the elements that are there. Just use the ones that you want.

There is inevitably a certain amount of chopping and changing. This isn't a problem. This is your mind thinking through how it will use this information. When you come to write or speak about these elements, the time you spend organising them will pay you back.

Go through this process with every strand that you've chosen (leave the leftovers and the high-impact strands for now).

At the end of this, two outcomes are very likely.

First, you will have some elements left over. Double-check that you don't need them. If you don't, move them into the 'not sure' strand. You might turn back to them later.

Second, as you begin to order the elements and you really start to engage with how you want to tell this story, you may spot a missing element that you really need. My BBC colleagues and I have come to call these 'shopping list items'. I use the phrase all the time now. When I'm writing scripts, talks or articles, I write 'SL' next to areas where I know something is lacking. I or one of my colleagues will then set about finding what's on the list and, if we can find what we need, we add it into the gap we've identified. A short shopping list is always welcome, but if it's long, that's OK too. This part of the process is really important to ensure both that you are being comprehensive *and* that your story flows as you would want it to. It's this attention to detail that can elevate the work that you're doing.

By this point, this is what you should have:

- a series of subject strands with ordered information
- some leftover elements and elements you're not sure how to use
- some high-impact elements you're still waiting to use

THE HIGH-IMPACT ELEMENTS

How do you imagine starting this whole story off? Which elements could help you end the story? Select the ones you think could perform these roles. Place them together in new strands marked 'Introduction' and 'End'. What you select will be dictated by the story you've chosen to tell, so keep this in mind. You may need quite different elements at the start and finish depending on how you are going to tell the story. There's not a right or wrong answer.

VISUAL ELEMENTS

Now that your information has been distilled and organised, if the nature of your explanation allows it, it's worth thinking about which visual elements you could use to give extra impact and clarity.

Look at each strand and go through the information you have in each one.

- Are there phrases or facts you want to emphasise, a series of actions you could highlight one at a time, sections of your explanation that you want to mark as you work through them? Are there graphics, maps or images that you want to show?
- What do you plan to show at the start?
- What will be the final image that remains visible when you finish?

It's hard to overstate how much difference these visual elements can make if you get them right. It's also hard to overstate, as I was discussing earlier, how distracting they can be if you get them wrong. To avoid this trap, select visual elements that explicitly support what you're saying. If you want to quote someone, pull up the quote. If you want to show an event, show a picture of it. If you want to reference a statistic, show the statistic.

Try to avoid generic visual elements. If you are talking about cars, I'd avoid a random image of cars. If you mention France, you don't need the Eiffel Tower. Now, if you were talking about a specific car – then, yes, show it. 'This car has changed our industry' connects your words to the image. 'The number of cars being bought is falling' while you show a generic image of some cars does not. A graph would, though.

At best, generic images do nothing; at worst, they distract and create a sense that you've nothing of great value to show.

Start pulling your visual elements together in the order you think you want to use them. This isn't a fixed selection. You can add to this or leave some of these elements out. What's important is to start assembling the visual elements you may need and knowing which pieces of information they connect to.

In Step Seven, we'll look at how to really focus on the way you use these visual elements but, for now, we just need to start collecting them.

QUICK CHECK

- What is the purpose of your explanation?
- What is the story you want to tell?
- How do you want to tell that story?
- Do your strands work?
- Check the list of things that you need to understand better.
- Are you missing any visual elements that you'd like?
- Have you discovered new areas you think you need to cover?
- How do you feel about the information in your 'not sure' strand?
- Do you need advice?

Again, some of these questions are familiar territory. Some won't take more than a moment to answer. But the discipline of checking as you construct your explanation is invaluable at ensuring you're on track.

That brings us to the end of Step Four. A lot of hard work has gone into getting this far. Now comes the reward. For me, this is where the fun begins. We've taken great care to assemble, refine and organise the information we think we need for our explanation. Now we need to turn it into something that is coherent, consumable and, we hope, starting to resemble a great explanation.

STEP FIVE:
LINK THE INFORMATION

This is where we start to tell our story. If you're producing a written explanation such as an essay or report, we're going to start writing around the elements. If you're producing something to be spoken, such as a presentation to your boss or a conference lecture, you can either write yourself a guide script or experiment with talking through what you are going to say.

For the sake of this section, we'll assume you are going to write down your explanation. After we've done that, this can easily be adapted to use as a spoken explanation.

If you are preparing a lecture or a presentation – something spoken but where you are in control – there will be times when you do want to read precisely from a script so that you can be in absolute command of everything you utter. In which case, writing the whole thing makes sense. While writing this book, I gave a speech at a high-profile journalism conference where I knew some of what I said could be contentious or misconstrued. In that case, I wrote a full script, checked with my editor and stuck to it.

For most of us, though, those moments are the exception not the rule. Even in other cases of public speaking – a presentation in a conference panel or a talk to colleagues or students – I'd certainly advise planning what you want to say but strongly advise against regurgitating the whole thing word for word. You'll find much more later in the book about speaking with clarity and purpose either with guide notes or no notes at all. Both can be much more effective at explaining yourself than simply reading something out. For now, we're going to write around our elements. Even if we don't end up reading every word out, this will help create a familiarity with the structure and the phrases that we want to use.

As you prepare to write, keep these questions in mind:

- **Is the language you're using as simple as it can be?**

- **Are you clear what role each element is playing?**

- **Are you sure precisely what you're trying to say with each sentence?**

- **Are there areas you still don't understand?**
  If there are, jot them down.

- **Do you have your list of the questions that you think people will have?**
  Keep it close by. As you reach sections where those questions apply, make sure that you have answered them. And if you can't or don't answer them, that's one for the list of areas you still don't understand.

By this stage of the book, you'll have noticed I'm asking some questions repeatedly. This is because they need to be ever-present when you're seeking to explain anything well. In time, my hope is that for every explanation you take on, you'll automatically be asking these questions. But if this is a new way of doing things, then, just as with anything we learn, making it systematic at first can help. That's why I'm asking you to come back to these questions several times over.

LET THE WRITING BEGIN

I always recommend starting at the beginning. Humans tell stories from the start and I think you'll find it easier to shape the story you want to tell if that's where you begin. Take each section of your structure in turn and start each section with the element you've placed at the top.

As much as possible, let the order that you've given the elements guide you – and let the elements themselves guide you. Remember you've chosen each one to serve a purpose. Let the words you choose explain that purpose. If your structure is working well, you may find this flows reasonably easily. I remember when I originally devised this system to help me write history essays at university, I would spend up to two days getting to this point. By the time I came to write, it was a case of stitching everything together in under a day. It wasn't easy but a lot of the hardest work had been done and on a good day the writing almost looked after itself. The same can be true on a smaller scale: I may spend an hour getting to this stage of the process on a briefing paper and then write it in fifteen minutes.

As you get going, if it's not flowing, don't panic. This is completely normal. You could try starting the section with a different element first. Maybe that will feel easier. If you're still stuck, try to identify why. Is it because you're not sure exactly what you want to say? Or you're not sure how to express it? Often when I hit a block, it's because I haven't actually settled these two questions. There may also be a practical problem. Do you need to discard some of your elements? Do you need some extras? Or maybe the order isn't quite right. It's very unusual that the order you originally chose is precisely what you end up with. Some movement is to be expected.

Just as important as all those questions is to keep going. I frequently will force myself to start writing even if I'm not feeling sure and I'm not happy with what I've written. I may then rewrite what I have multiple times. Bit by bit, I'll edge towards something that works. Much like this whole process, even when the writing is feeling uncomfortable, the best thing you can do is start. The process of trying and failing will increase the chances of you getting it right.

WRITING TECHNIQUES TO HELP YOU EXPLAIN *AND* TELL A STORY

Just as I have a range of storytelling models to which I will turn to give shape to a particular explanation, I've also got a number of writing techniques that I use to give my explanation clarity, impact and, I hope, narrative momentum too. Some I use all the time, some only when it feels helpful. Here is a selection:

Avoiding hard stops

As you add in sentences, have in mind ways to keep people moving from where you *are* on to where you are *going*. Radio and TV shows do this all the time. If you listen to music radio or twenty-four-hour news, you can see and hear the presenters always doing their best to avoid any form of 'hard stop'. This is a moment where what you're consuming reaches a conclusion and, to the broadcaster, this is a moment of danger – a moment people might switch off. To tackle this, they come up with lots of ways to say, 'It's worth staying'. I'm certain you'll be able to think of endless examples.

After the break, we'll play the new Adele track.

We're going to go live to North Carolina in a moment on the storm. And after that, we'll be hearing an update from the President on the ongoing situation.

This kind of technique will be familiar to you. In broadcasting it's called 'trailing'. It's aimed at people who've already started watching or listening and who you want to hold on to.

There are other examples from the world of TV. If you're watching live, at the end of a programme, as the credits roll, the continuity announcer will come on over the theme music and say, 'Coming up in a couple of minutes, a brand-new drama set in London in the sixties.' They used to do this after the credits but then they realised that the credits themselves were such a moment of danger and so decided to intervene earlier. Similarly, if you're watching on a streaming service, within moments of the credits starting, they're offering you the next episode or another programme entirely. All these services understand how much of a threat a hard stop is to their business. We need to see hard stops as an equal threat to our explanations.

You've already done plenty to support the success of your explanation by assembling essential information in a logical order. Trailing is another way of trying to maintain interest. You may have noticed me doing this repeatedly in this book.

Trailing

Here are some examples of trailing in the context of an explanation.

If you're wondering how we achieved that uplift in sales – we did three things. And I'm going to show you each of them.

I've talked about how voters are changing the criteria that decide how they vote. I am going to show you – and then afterwards, we'll look at how that connects to the way their news consumption habits are changing.

I want to take a few more minutes to tell you the reasons we decided to launch a new product. Once I've done that, I promise I'll show you it – we're excited and I'm pretty sure you will be too.

Trails take two primary forms. They either entirely look forward. Or like the first example above, they look back at something that's just been referenced – and then throw forwards. Both can work and both can be positioned between the 'subject' strands that you've built the structure of your explanation around.

Trailing is very effective at marking your overall direction of travel. But it can only be deployed at major junctions in your explanation and, as such, you're unlikely to use more than a few in any one piece of work.

Surfacing the structure

This is a technique that directly connects to trailing. I've already mentioned how trailing can signal further sections of your explanation that are to come. Surfacing the structure does this as explicitly as possible.

When I was studying A-level history, my teacher, Miss Thomas, would emphasise the need to signpost each section of my argument as I wrote an essay. It was excellent advice.

When we plan anything, from an article to an academic paper to an essay to a speech or a presentation, we always think about a structure. We've done that in this Seven-Step process. That structure helps us organise our thoughts, but it can also help the people you're addressing see the subject from the same vantage point as you. It can give them the same bearings. This technique simply involves not just having a structure but saying so. There are lots of ways of doing this.

Now we have looked at X, the next part of this issue is Y.

We can't understand X unless we consider how it connects to Y.

So far we have considered X and Y – but Z is also important.

Put X and Y together, and that leads you to the next factor: Z.

X, Y and Z are all important – but none of them matter as much as A, B and C and we're going to look at those now.

I've explained how X happening led to Y and how that in turn led to Z. That series of events meant that A became inevitable, as did B. Let me show you how.

You are telling the person you're speaking to what they've heard and what they're about to hear. That's always going to be helpful.

One variation on this is if you need to take your explanation on a tangent. All explanations have a start and a destination. Most are linear – and linear communication is easier to follow. However, on some subjects, being linear is not always possible. You may need to go at a tangent to provide necessary background or context. That in itself is not an issue but there is a greater risk people will struggle to follow you. If you 'surface the structure' you can help people stay with you and follow what you're doing. To put it another way, tell people you're going at a tangent and tell them when you're back from it. Here are some examples:

Now – I mentioned X. We can't go any further without examining how X fits into this issue.

It's at this point that we need to pause and consider X.

I'll pick up what happened next in a moment – but before that we need to talk about X for a couple of minutes.

Having done that you can return to the linear route you were taking.

Now, bearing all that in mind, let's turn back to where we'd got to.

That's X. We have to keep it in mind as we work through this. Now back to what I was saying about Y . . . [and you pick up the story].

Saying what you're doing can give you confidence as you deliver an explanation. It can be helpful to whoever's reading or listening, too – especially when your explanation is lengthy and/or complex. That clarity, we hope, reduces the chances of someone stopping reading or listening (sadly for Miss Thomas, an option not available to her as she marked my essays . . .).

Joining phrases

I'd long been tuned in to the idea of trailing. But in 2021 I had a realisation that changed how I approached explanation. We've established that trailing and surfacing the structure are ways of creating momentum through the big junctions – but what if, instead of doing this solely at those points, I tried to do it *all* the time?

Trailing was something I was doing occasionally between the sections of my explanations. In fact, I used it much more on the news. In between one story and another, we'll trail to other stories that are coming up. There are even places in most news programmes' running orders where the trails go. If you have a news programme you watch or listen to regularly, you'll know where they go. This isn't a problem, but what I'd noticed was that it separated the idea of *keeping the audience* from the telling of the stories. We trailed between stories and sometimes between sections of a story. I wanted to see if I could build how I held people's attention and interest into the fabric of my explanations.

To do that I started experimenting with what I call 'joining phrases' and 'hooks'.

Joining phrases take the person you're addressing from one thought or element to another. They escort the viewer or reader from one element to the next without, we hope, a chance to consider tuning out.

Here are some examples of joining phrases:

That fact helps us understand this issue. But it's not the full picture. To get that, you also need to factor in X . . .

My first goal had been achieved. But that immediately led me on to a second goal – which was to prove a lot harder.

If some colleagues were telling me they were angry at what I was suggesting, listen to what else I was being told.

Those were the numbers for last year. Now look at what has happened this year so far . . .

You may have noticed that these joining phrases look back and look forward in the same breath. They're micro-trails, if you like – instead of joining major sections they join elements. Our nuggets of information. They slingshot the person you're addressing from one important element to another. They are potent at creating momentum and also help to minimise 'hard stops' anywhere.

'Back annos' and 'hooks'

In our explainer videos we had already been experimenting with what we call 'back annos' – which is short for 'back announcement'. This is TV jargon for a script that comes off the back of something.

Here's a regular example of a 'back anno'. We play a clip of someone and directly after it, I say, 'And you can see that full interview on BBC1 at 9 p.m.' It's a trail that connects to whatever it's following on from. Initially I didn't see that this technique could help me with my explanations. A trail of that nature comes off the back of one story and leads to a hard stop. I pause and then move on to the next story. But I started to see possibilities if I used back annos in a different way. Instead of using them as basic trails, I started to use them to react to and give emphasis to what we'd just heard.

To do this we first need to identify what we want to emphasise. This might be a phrase, statistic or single word that is particularly important. I call this the 'hook'. It's whatever I can get hold of as I come off the element.

Back anno for emphasis

This almost always involves repetition. Before moving on from an element, you repeat what we've just heard for emphasis.

Let's imagine we have a quote from the head of police.

At this stage, we can't be sure when approval for this operation was given.

I could back anno this: 'And while the head of the police says he *"can't be sure"* how this happened, the families of those affected say they know.'

Or here's another example of a back anno:

We need to be clear what we're facing. A recent assessment found that revenue from this sector of our industry had fallen 90 per cent in two years.

Ninety per cent. In two years. When the industry as a whole has grown 30 per cent.

If you can spot potential hooks, they can really give impact to a piece of information. It can work in written or spoken explanation. Just remember it's a trump card. If you overplay it, it will lose its sting. Save it for when it really matters.

Hooks and back annos can also do more than add emphasis.

Back anno to add context

Here's an example of how it works. Let's imagine this quote from a fictional President who says: 'That's why today's announcement of a new climate change goal is a significant step for our country.'

I could back anno that in this way:

It may prove to be a significant step, but at the moment the President hasn't provided any detail on the actions that will deliver this.

I want to highlight that the policy hasn't been fleshed out. The 'hook' is the words 'significant step'. That is the primary claim the President is making. I want to emphasise the phrase as a means of emphasising both the goal *and* the lack of information on how it will be achieved. In this back anno, I am using that hook to directly engage with what we've just heard and to introduce a necessary piece of context.

Here's another example (with fictional statistics):

The head of one industry body says that 'the amount of money being spent on vintage clothes is now £100 million.'

£100 million sounds like a lot. But the country spends £10 billion on clothes every year. Vintage is growing but it's still dwarfed by the sale of new clothes.

Once more I use the hook for emphasis and the back anno for context.

Joining phrases with a hook

Put all this together and you have a technique that allows you to add impact, add context if need be and to transition through to the next element of your explanation. Here we want to spot a word or phrase that will be our hook, we'll then repeat it and use that thought to throw us forwards. Let's look at an example.

> *Our suppliers have found this new approach hard. One told me: 'I just don't see why this is necessary.' But while they don't think this is necessary* [the hook] *– all the data on our supply chain suggests it is* [the joining phrase]. *Look at this from our industry body* [the next element you want to introduce] . . . [go into chart showing the latest data].

This leaves us with some important additions to how we're communicating.

Here's a classic way to explain something.

1. Introduction to some information
2. Share the information
3. Introduction to some more information
4. Share the information

We move forward without looking back. Here's how we can adapt it.

1. Introduction to some information
2. Share the information
3. *Back anno to add context*
4. Introduction to some more information
5. Share the information
6. *Joining phrase with a hook*
7. Introduction to some more information
8. Share the information

In a short space of time, we can inject momentum, context and impact.

The biggest shift I've made in how I explain subjects in recent years is to think about creating momentum all the time rather than only in the main junctions within an explanation. It's been transformational. Think of it as constant velocity rather than attempts at acceleration. To do that, keep in mind where you're coming from as well as where you're going. Look back and look forward . . .

Parallel chronologies

If I tell you this one, you'll spot me doing it over and over again. I know of few techniques that are as effective at maintaining momentum and engagement. You'll be familiar with it from novels, non-fiction books, films, sport coverage and documentaries. Here, you tell one part of the story and, before you go any further, you tell related parts of the story that have happened elsewhere.

If you're a football fan, you'll get this every Saturday afternoon and especially on the last day of the season. The presenter will say, 'With that goal Manchester City are now in a position to win the league, but, hold on, a goal has gone in at Liverpool that changes everything.'

If you're watching a politics documentary, the narrator may say, 'But while the President was meeting his allies, across town, in the offices of one of the rebels, the plot to drive him from office was taking place.'

Many stories have parallel strands and the same is true of explanations. If you can move between them, creating a sense of different events playing out in parallel, it can be a powerful storytelling device. As well as creating narrative momentum, it's very effective at juxtaposing competing developments to highlight how they contrast and interlock.

For example, in a natural history documentary, the narrator

may say, 'The efforts to reintroduce the native oysters to the river had been a success. But as their numbers multiplied, a new threat was emerging – algae.'

'While' and 'as' are your friends when trying to create this effect. With 'meanwhile' waiting in the wings if need be. Here are a few versions of these kinds of joining phrases that help you do this.

> *X had done well. But while we were pleased with that, it was clear Y was still an issue.*
>
> *As X was happening in one place, Y was happening in another.*
>
> *While X, Y and Z were resolved – it was becoming clear A and B were not.*
>
> *So in one office I was being told X, while in the other I was being told Y.*

And on they go. These phrases can help you shift from one strand of your explanation to another, they avoid any sense of a 'hard stop', they create momentum and they connect strands together.

If you want other examples, I'd be surprised if you could watch any three of my videos and not find some.

Splitting sentences into two

'There's nothing wrong with the content but I've split the sentence in two.' The brilliant producers who work with me at the BBC will be very familiar with this turn of phrase. I say it an awful lot. In Step Six we're going to look at how to 'tighten' explanations. I do this with every story that we put on air and probably the most frequent change I make is to turn one sentence into two. (This isn't a criticism of my colleagues – I frequently do this to my own writing. When all of us first write, more likely than not we won't serve up perfectly formed short sentences.)

This is particularly important if you are going to be sharing your explanation verbally. Long sentences are simply harder to take in than short ones – especially when we are hearing them rather than reading them. They're also more difficult to deliver because organising your breathing is harder.

For any verbal explanation, I have a rule of thumb that I don't want to go beyond fifteen words. I'm not rigid about this but it's a helpful guide. And I try to keep sentences short in all circumstances. Here's a simple example of something I might change whether I was writing it or saying it out loud.

Original text

Swimming lessons are open to children of all ages – we have eight stages which are available through the week, and you can sign up your child on our website or at the reception which is open from 9 a.m. to 5 p.m.

Edited text

Swimming lessons are open to all ages under sixteen.
There are eight stages in total.
Lessons take place every day, including weekends.
Register on our website or at reception between 9 a.m. and 5 p.m.

Now, I'm not going to jump the gun on the tightening process here, but, needless to say, you'll have fewer changes to make if your sentences are short and easy to consume from the start.

Put the subject of the sentence near the front

There is a habit in many forms of communication – especially spoken – that refuses to go away. You'll often hear the subject of a sentence placed a long way into it. Here's a (fictional) example of what I mean:

Low on MPs, losing support, and surrounded by controversy, the opposition is under serious pressure.

If this is the first time that we've mentioned the opposition, the whole time we're listening to the list of problems we're thinking, 'Who is this about?' And if we're thinking that, we're not concentrating on the list.

There's a famous example of this from a few years ago when a BBC presenter said:

This is BBC World News. I'm Jonathan Charles. Kept hidden for decades and forced to bear children . . .

The subject of the story is about to follow but it sounds as if the subject is Jonathan himself, which is why it's been watched millions of times on YouTube. Believe me, we have all read scripts like this.

There's a twist here, though. Jonathan's now a strategic communications consultant (and so knows a huge amount about effective explanation, just as he did when he was a presenter). I contacted him to ask about the clip. It turns out it's not quite what it seems. The clip is from a rehearsal where Jonathan was quickly reading through the script. He tells me, 'When it was broadcast, it worked because it was slower and covered by pictures of the woman concerned.'

I fear that detail has been lost on YouTube's viewers. But the clip and Jonathan's explanation of it teach us a number of lessons. First, how we deliver words affects their meaning and what we are showing when we say words affects their meaning.

There is a more direct lesson too: that if you place the subject of your sentence in the middle of it, you're taking a risk. In part because the delivery has to be right for it to work – but also because the audience may be left wondering what the subject is.

I'm always inclined to lower the risk of miscommunication – so for me this is a simple but helpful rule: reference the subject *before* you talk about it.

The power of 'and'

This technique is definitely better suited to verbal explanation than written explanation. I know that because I am wrestling with this right now as I write this book. To my disappointment, my heavy *'and'* habit does not easily translate into written explanation. It's still worth mentioning, though.

Let's consider some of the things we're trying to achieve as we explain ourselves: momentum, the sense of one element connecting to another, the sense of constructing an explanation block by block, the sense of leading somewhere – and we want short sentences too!

The word 'and' at the start of a sentence can help you with all of those. 'And' is saying to your audience 'that's not all' or 'I've got more for you'. It's asking them to come with you. It's the very shortest of joining phrases. Here are some examples of how I might use it. The first is an adaptation of one of our video scripts:

Example 1

The government has just announced further funding to insulate the least efficient homes.

And this investment is part of a broader energy-saving strategy.

The second sentence looks back at what I've just said and then introduces a new point.

Here are some other fictional examples:

Example 2

The weather had been bad. The attendance had been bad. The ticket revenue had been bad.

And to make matters worse, many of our staff were struck by a bug.

Example 3

The challenge had always been to have some colleagues working from home and others in the office.

And while we were figuring that out, the chairman and the CEO resigned.

Try reading those three examples without the 'and'. That would be the more classic way of writing them and there's nothing wrong with that. But when you add the 'and', can you feel how, when you read it or say it, you flow through the two sentences better? It gives it a rhythm and momentum that is harder to achieve without it.

I use 'and' multiple times during any spoken explanation. In writing, it can sometimes jar a little depending on the style. I do still use it (as you'll have seen in this book) but nowhere near as much. That is one for you to gauge – but when speaking it can really help.

DOES THIS SOUND LIKE ME?

A year or two back one colleague came over to speak to me. She very generously enthused about our programme and, in particular, its tone.

'I just don't know how you fill a whole hour without scripts,' she added.

'That's very kind but I don't,' I replied, thanking her. 'It's *all* scripted.' And it is. What I'm aiming for is something that has the precision and accuracy of a classic news script but also feels like me talking to you – just as I might do if we met in person or if I came to give a talk and you were in the audience. And how I go about this connects to my early experiences of going on air.

Dotun Adebayo was a fixture on the 5 Live show *Up All Night* when I was offered my first BBC presenting shift. The show had been created years before by another presenter called Rhod Sharp whose warm, lyrical style we all loved and was very much the original sound of the show. Rhod and Dotun now shared out the week's responsibilities and, with my first shift looming, I asked Dotun for his advice. 'Whatever you do,' Dotun told me, 'don't try to sound like Rhod!' He wasn't being rude – far from it. We were all fans. Dotun's point was that Rhod's inimitable style was his and his alone. Anyone else doing the show would need to play to their own strengths and not do an impression of Rhod.

This rang true. Partly, because I could hear Dotun very successfully doing his own thing. Partly, because I thought back to the first time I'd ever presented on live radio. This was during the time when I was unemployed and had gone to the *Independent* for that ill-fated meeting. During that summer, the one thing I'd got going was a midnight-till-2 a.m. slot on an FM station with an audience that, I suspect, could have been measured in the tens. While trying to persuade the programme manager to let me and my record collection loose, I had made the case that I'd bring a more authentic, less stylised form of presentation than the classic daytime output. 'You say that,' he replied drily, 'but you'll be trying to sound like Gary Davies before you know it.' (Gary Davies was one of the biggest presenters on BBC Radio 1 in the eighties. I was an avid listener and a big fan, and Gary's still going strong on BBC Radio 2 now.) He wasn't being rude about Gary Davies or any of the big mainstream radio presenters. His point was that it's human nature to copy others who are doing what we're about to try – and that it's harder than you might imagine to escape that. Sure enough, not long into my first show I was telling listeners to 'stay locked to 107.9FM' and so on. Finding my own voice, it turned out, was much harder than I imagined (as indeed was radio DJing all round). And that problem wasn't a one-off.

Going back over old tapes of some of my early forays on the news, I can hear the same thing. Unsure of how to be, I am aping those who are already there. In terms of the basic presentation techniques I needed, that was definitely worth doing. But in terms of establishing my own style and tone, much less so. It turned out one simple rule would unlock this.

On every single TV script, every video, every presentation, every interview I prepare for – I ask myself: *Would I talk like this?* If the answer is no, it has to change. No ifs or buts.

I know how I talk. You know how you talk. You know what you wouldn't say. You know what you would say. Let all of this guide you.

This single question informs how I write scripts for verbal explanations. It also influences how I write articles, essays and, yes, even a book. Each medium may require you to modify how you express yourself – that's normal. We all speak differently at a party to how we speak in a meeting or in a presentation, but there's never a need to trade in who you are or how you speak.

I hope, for better or worse, if you were to meet me or hear me talk after reading this book, there wouldn't be a difference.

BRING THINGS TOGETHER

This is a quick and effective technique to either conclude your explanation or a section of it. As I approach a conclusion, I like to mark that what I am about to say is the result of what has gone before. When I was writing essays at school, I used to write 'in conclusion'. It does the job, but I much prefer 'all of which'. It explicitly says that everything that's gone before leads to this.

Here is an example of how we could end one strand and move to another:

If you are explaining the start of the First World War and you have listed how the war played out in its first few months, you could then write:

> *All of which meant that by the winter of 1914, the war had already reached a stalemate and the Allies were ready to seek a new strategy.*

In this sentence, we signal that everything so far has fed into this moment and then we throw forward to what comes next.

Or here are some examples of completing a whole explanation:

> *All of which means that our current business model has, sadly, reached the end of the road – but also that an exciting alternative is emerging.*

> *All of which means that there is a once-in-a-generation opportunity to reshape how politics works – and a passionate discussion of exactly what form that reshaping should take.*

> *All of which means I think I'm in the best position to take on this role.*

If you don't want to use 'all of which' all the time – and we should definitely be alive to not sounding repetitive – here are some alternatives.

> *And so if we consider all of that . . .*

> *If we consider all of these factors . . .*

Put all of those developments together and . . .

If all of this is combined, we start to understand why . . .

You'll be able to come up with plenty of variations of your own. But phrases that signal you're about to conclude and set you up to do just that are very useful in explanations. They help package the information for the consumer and explain where it is leading us to.

Thinking about your conclusion is always time well spent. However well we communicate, not unreasonably whoever we're addressing may not take in *all* of what we're trying to get across. A combination of both a conclusion and a reminder of what we've heard can be both helpful and add extra impact.

The point that has been reached

This is one more way to bring things together. I've split it out from the rest above because it's doing a slightly different job. If the phrases in the previous section were setting up a conclusion, this technique sets up a summary. This will take up more time, so it won't always work. But if you have the space, in the same way a back anno with repetition can provide emphasis, so a summary can do the same thing for your entire explanation. The phrase I often use to set this up is: 'the point that has been reached'. An alternative is: 'this, then, is what we'd done' or 'this, then, is what has happened'. It signals that I am going to tell you where we've got to and/or how we got there. Again, it can be used to end a section of your explanation before moving on. Or it can be used at the end of the whole thing.

To do this effectively, you need a list of the aspects of the point that's been reached. This can be heavily influenced by the strands and essential information that you've already assembled. This summary is a reiteration for emphasis; it's not the introduction of new information.

This is how it can work.

If you were using this to join two parts of your explanation, this is how you could do this.

> *And so after three years of working on the idea, this is the point I'd reached . . .*
> *— Fact 1 (each fact being a key characteristic of this moment)*
> *— Fact 2*
> *— Fact 3*
> *As you can see, the situation wasn't viable. That was when I decided to call X . . .*

If you were using this at very end of your explanation:

> *After three years of war, this was the point that had been reached . . .*
> *— Fact 1*
> *— Fact 2*
> *— Fact 3*
> *— And a concluding thought: . . . All of which meant the efforts of the Allies to force a swift end to the war had failed.*

Here are two more:

> *The policy has been in place for twelve months now and this is the point we've reached with it* . . . [List the outcomes of the policy and the assessment of it.]

> *The internet has been disrupting our business since the nineties. We've heard how. And now, in 2022, this is the point that we've reached* . . . [List the core facets of the situation the business is in.]

You get the idea. Here are two examples of me doing this during the US election of 2020. The first is from two days after the election when Donald Trump was refusing to accept the result.

It's another day for America's democracy. The President's making baseless accusations of electoral fraud and attempts to steal the election. His son Eric is pushing misinformation about ballot burning when there's no evidence that that is happening. His campaign is launching multiple lawsuits to stop vote counting when the counting of votes after election day is a standard part of the electoral process. And international election observers are accusing the President of a 'gross abuse of office'. Now, Donald Trump repeatedly attacked America's democratic process during the campaign so the last two days are not entirely a surprise. But reaction around the world shows us the President still has the capacity to shock. For better or for worse, this is not American democracy as people normally see it.

By listing the elements that made up the moment there is a cumulative effect.

Another, more farcical moment, came days later when the Trump campaign set up in the yard of a gardening firm called Four Seasons Total Landscaping in Philadelphia. It appeared they'd meant to book the Four Seasons Hotel but had made a mistake and gone with it. The gardening firm was next door to an adult book store called Fantasy Island and as the event in the firm's yard went on, Joe Biden's victory was announced by one news network and Donald Trump was filmed playing a round of golf. The report we made finished like this:

And so at 11.24 on Saturday, America reached a historic moment. CNN announced Joe Biden as the President-elect,

Donald Trump played golf and his campaign was in a car park by Fantasy Island.

'It's another day for America's democracy' and 'America reached a historic moment' are simply variations on 'this is the point that that's been reached'.

It's a technique that can help reinforce what you've already explained.

THE BEGINNING AND THE END

Your first and last sentences are your best chance of laying out what you hope to explain and the conclusion you've reached. Get them right and you give people a strong reason to keep listening and you leave them with a thought that may stay with them.

These are the sentences I rewrite the most. They are also often the ones I settle on last. When you're writing at this stage, try to capture what you'd like to get across at the start and finish. It's much better to have something than nothing. Don't worry if it's not the perfect form of words yet. You can keep coming back to it. I find the more I work on an explanation, the deeper my understanding of the subject. With that understanding comes a clearer view of what matters most, which, in turn, can guide how I start and finish. Sometimes, it's only right at the end of the process that I see clearly exactly how I want to do that. What's important, though, is that you're mindful of this throughout.

I'm not enthusiastic about easing into explanations. From the start I'm looking to pique the interest of those I'm addressing and draw them into the story I want to tell. Then to finish, I'm looking for a sentence that allows me to apply emphasis to my explanation.

EXAMPLE

These first and last lines are from a video explainer that we made about the tensions surrounding the self-governing territory Taiwan.

The start:

> This is the story of how a historic dispute has become part of a global power struggle.

Here, I'm attempting to make the issue feel relevant to everyone even though some viewers may feel Taiwan doesn't connect to their lives. I'm also trying to emphasise the urgency and importance with the phrase 'global power struggle'.

Then this is how I finish:

> And so Taiwan's status has become a test of the limits of America's power and the possibilities of China's.

Here I am re-emphasising the twenty-first-century context of this issue. The idea I want to communicate is that this is part of a fundamental recalibration of global power because of China's rise. It's a thought that makes this explanation about Taiwan's particular situation but also makes it about the evolution of the world as we know it.

Start well and there's a good chance people will stay with you. Finish well and there's a good chance they'll think it was worth it.

HIGH-IMPACT PHRASES

In any explanation, we need phrases that can encapsulate what we're trying to say. These are the moments where we can cut through and really communicate. Every element of our explanation matters but, on top of that foundation, high-impact phrases give us a much better chance of lodging our core messages in the mind of our audience.

It's always worth being conscious about whether you have these phrases in your explanation. These are especially useful if you have to explain something multiple times.

I find high-impact phrases come to me in three ways.

The first is I work hard on them!

Back in the early 2010s, when I was pitching *Outside Source* as a new format to the BBC, I had Word documents stretching to pages and pages with all the ideas and strategies I hoped we could pursue. It was useful for me to collate everything in one place. It wasn't useful as a way of selling the idea.

I boiled those pages of notes into a single-sentence summary. I wanted *Outside Source* to be 'a hyper-efficient authentic global news bulletin containing only on-the-day stories that use all the best BBC and non-BBC content in whatever form it comes'.

As a way of defining the ambition of the programme, this might be a useful guide for the bosses and my colleagues in the newsroom if and when we came to make the programme. Indeed, it has been – I've referred to that sentence plenty of times over the years. But in the early days, when I was trying to get the idea off the ground, this wasn't the high-impact phrase I needed. For a start, it was too long and had a number of ideas within it. It was a lot to take in in the middle of a conversation or a short presentation. Instead, I worked on much shorter phrases that I could say. These are some examples:

A TV news bulletin for a digital world

A broadcast version of a livepage

Real-time collation

We can construct stories in front of the audience

Informal and informed

We'll prioritise depth and context – it's news for people who know the news.

There were others too. These high-impact phrases were at my disposal. I didn't use all of them every time I talked about this idea – but I did use them an awful lot. I had more detail if people wanted it, but these phrases quickly and effectively got across the idea.

The second way that these phrases come to me is to notice when I say something that has had impact. When I was setting up the 50:50 Project, I had many meetings with teams who were interested in joining. The conversations often followed similar paths because, quite reasonably, a lot of my colleagues had similar questions and concerns. I started to notice the phrases that clearly helped me communicate my answers best. I'd note them and use them again. Equally, if I said something that clearly didn't help, I'd note that too. Over time, I was discovering which phrases were most effective at giving my explanations impact.

The third way I find high-impact phrases is to listen for them when others are speaking. My explainer videos have become associated with the phrase 'assertive impartiality'. It's a really effective way of summarising what I'm trying to do. It's not mine, though! As I mentioned earlier, a former BBC executive called Gavin Allen used it in an interview when referencing my work. People immediately picked it up and would ask me about it. It clearly captured what I was doing and caught people's imagination. I started to use it straight away (with attribution of course . . .).

For me this is such an important lesson. Sometimes others capture what we're trying to explain better than we do. We can lean on them to give our explanations clarity and impact.

However you find them, high-impact phrases are an incredibly powerful tool.

COMPLETING THE FIRST DRAFT

All the techniques I've suggested can add to your arsenal. Use them when you think they work, not because you feel you ought to. Hopefully they'll help as you turn your strands and elements into something that can be read or heard.

By this stage, you'll have a draft of your explanation. This is a really important moment. The work isn't done but you've got something tangible now. When I got to this stage writing this book, I was under no illusions that it was close to being done. But for the first time, I felt it was possible. Those feelings of possibilities can deliver a much-needed energy boost as we push on to the finish line.

However, before we complete Step Five, there's one more thing I'd like you to do.

GO FROM THE TOP TO THE BOTTOM

This is a first chance to see how your explanation feels from start to finish. We're not worried about everything being perfect at this stage, but we do want to check if it's hanging together.

- Are you happy with how the strands and elements move from one to the next?

- Are there gaps in your explanation?
- Are there areas where you're struggling to get your point across?

In all cases, you can either take action there and then or at least mark that more work needs to be done. Either way, we want to catch any areas where there's more to do.

QUICK CHECK
- Are you happy that you've stuck to your story structure?
- Have you used different techniques to provide emphasis and momentum?
- Does it sound like you?

You'll be relieved to know this brings you to the end of Step Five. We've come a long way from that moment at the start where the subject and the volume of information was considerable and, perhaps, felt overwhelming. Now, we're into the edit.

STEP SIX: TIGHTEN

This is the moment where we can turn a good explanation into an exceptional one. Doing this reminds me of the dark, often dank mornings before school when I'd do my piano practice. Even if I had roughly worked out how to play a piece, I'd still have to put in the hours seeing if I could refine each section and create something that worked as a performance. Without that, I'd have a few notes I could just about play but it didn't add up to anything coherent and complete.

Always for me, Step Six is where I give myself a shot at something that really works as an explanation. Without it, we risk having some vital information that doesn't add up to the sum of its parts – and which isn't as consumable and compelling as it could be.

The late French music producer Philippe Zdar captured this moment. In the middle of a long YouTube video all about how he worked with the band Phoenix, he discusses how he mixes a track. He describes how he sets all the levels of all the different elements when the creative process is all but complete and the group is settled on what it's trying to achieve.

'We got to mix when it is really the end . . . It's very hard for the nerves because you have to keep your balance and cold blood.'

I allowed myself a wry smile when I heard him say this. He's talking about music not explanation but he's not wrong. Cold blood is required! It *is* very hard on the nerves. The process of tightening an explanation, just like producing and mixing a track, requires clear and, sometimes, ruthless decision-making. I say ruthless because you may well have to ditch elements that you've worked hard on.

Back in 2016, I spent a week visiting Silicon Valley. I met academics, journalists, media executives and tech leaders, and one idea came through consistently – the need to be able to cut something you're heavily invested in.

On a visit to 'X', Google's spin-off that focuses on long-term innovation, they went even further. One executive described how they'd throw parties to celebrate closing down a project that hadn't worked. The aim being to encourage staff to invest time and effort into ideas and then, having done so, to be able to divorce that investment from the decision over whether to continue. This runs against human nature. We tend to side with the things we've invested in (a phenomenon known as the 'sunk cost fallacy'). These parties were a way of trying to create a culture that could acknowledge and embrace failure rather than see it as a negative. No doubt Google's deep pockets make this approach a little easier, but it resonated with me, nonetheless.

The concept is found in newsrooms in a different form. The nature of daily journalism is that you're constantly bringing together work that has taken hours, sometimes days, sometimes weeks – and trying to offer a mix of stories at a particular moment, whether in a programme, a website front page, a social media feed or a paper. The nature of daily journalism also means stories get ditched all the time as new ones arrive.

This is particularly extreme in broadcast. I remember when I joined the BBC in 2001 being quite shocked and affronted when an hour-long discussion that I had set up was junked at very short notice because a story had broken. The editor (rightly) didn't give it a second thought. I had yet to learn the necessary degree of 'cold blood'.

This detachment is very useful when constructing an explanation. When we make our explainer videos, when it comes to the tightening, we all know the phrase 'I'm sorry I think it needs to go'. If the only argument against doing so is how long we've spent on it or how pleased I am with a certain turn of phrase, it goes.

Sometimes what stays or goes is a big decision. You will, though, be making lots of smaller decisions too. We want to refine each of your sentences to make them as precise and clear as possible. These smaller decisions are just as, if not more, important than dropping a whole section. The smaller edits can give your whole explanation efficiency and momentum.

In that spirit – and with your scalpel ready – let's turn back to what you have: a first draft which is ready to be tightened and polished. We'll take it sentence by sentence. We'll also not be afraid to ditch whole sections if they fail to deliver. As with every aspect of this process, seeing it all written down may make it feel like harder work than it is. Much of this will, over time, become automatic. You'll know what to spot. Not for the first time, as you start the tightening, here's a list for you.

1. Are there any obstacles to understanding?

Here we go right back to Allan Little and his obstacles to understanding.

Working from the start, how many can you spot? These questions will help.

Which personal or place names do you need?
Sometimes a name is essential, but often it isn't.
e.g. If you want to reference a US university, do you need to say it's in Texas?

Which dates do you need?
We often include more detail on a date than is necessary.
e.g. If something happened in 2012, does it matter that it's 9 May 2012?

Which statistics do you need?

Are you including multiple statistics to make the same point? Are the statistics you have directly relevant to the point you're making?

e.g. If the ruling party took 40% of the vote, the opposition 35%, the Green Party 10%, the Socialist Party 3%, the National Party 2% and 1% of voters spoiled their ballots, do you need to say how many voters spoiled their ballots? Or mention some of the smaller parties? That will depend on what you're explaining.

More broadly, have you included any information that is not essential to your explanation?

If you have, see how it feels with it removed.

2. Is there unnecessary complication?

See if any phrases could be shorter and simpler.

3. Could any sentences be made shorter without losing their content and meaning?

The key here is to constantly be looking for either words that are doing nothing or phrases that can be abbreviated without impacting the meaning.

Here are a few examples to give you an idea of what I'm talking about:

Before *That's smaller compared to the other one.*
After *That's smaller than the other one.*

Before *With regards to the offence that had been committed . . .*
After *As for the offence . . .*

Before *A player who had previously been suspended.*
After *A player who'd been suspended before.*

Before *Sales have seen a remarkable jump and we're all very hopeful that will continue.*
After *Sales are up significantly.*

Before *It can be argued that the US was left with no choice once China had taken its decision.*
After *Given China's decision, arguably, the US had to act.*

EXAMPLE

We've already considered information that, while interesting, is not working towards our overall purpose. As well as that, there are any number of words that find their way into our sentences that are serving no purpose at all.

Dr Richard Nordquist is professor emeritus of rhetoric and English at Georgia Southern University. I can highly recommend his list of 'Common Redundancies in the English Language'.[9] In his introduction, Dr Nordquist writes: 'Because we so often see and hear redundancies (such as "free gifts" and "foreign imports"), they can be easy to overlook.' He calls on us to 'be ready to eliminate expressions that add nothing to what's been said' and which make 'writing longer, not better'. And in a phrase that I love, Dr Nordquist describes how 'the phrases weigh down our writing with unnecessary words'.

What follows is a long list including:

- (all-time) record
- collaborate (together)
- during (the course of)
- gather (together)
- postpone (until later)

You get the idea. There are so many examples, both from Dr Nordquist and beyond.

With each sentence, stop and check if there are phrases that could be shorter or if there are words that are doing nothing. This, inevitably, leads you to shorter sentences. Here's just one example.

Before
Let's turn to a new flashpoint – between the Turkish military and the Syrian Democratic forces – a rebel group which opposes the government in Damascus, and controls much of the north-east of the country.

After
There have been clashes between the Turkish military and rebels who control much of the north-east of Syria.

The cumulative effect of doing this work is considerable. If every sentence is tightened, across even a short written or verbal explanation, you both give yourself extra time and increase the clarity of what you're saying.

---

4. Is there anything that is unexplained and could distract?

This is where we're looking for any reference to something that may not be understood. Think NATO . . .

5. Does what you're showing match what you're saying?

Are you happy that there is a sync between what people see and what you're saying to them? Do the words match the visuals? Are there any distractions? Do you give away what is coming next?

6. Does the start hook you in and the finish leave you with a clear conclusion?

Do the two complement each other? Are there any ways you could improve the clarity and impact?

7. Has your explanation answered all the questions that people will have about this subject?

Just as importantly, are you satisfied you've understood the aspects where you wanted to know more? And has that translated into your explanation? All being well, you'll have a satisfying moment where you realise you've addressed everything on your list.

8. And the big one: are all the strands and elements essential?

I know we've asked this a lot but it's not too late to ask again. If yes, great. But be open to the possibility that the answer is no. If it could be no, how would that section work without certain elements? How would the explanation work without an entire strand? If you're not sure, quickly try it. Often, you'll see almost immediately if it will work or not.

As you tighten your explanation, it'll gain clarity and lose distractions. It will almost certainly become shorter as you increase the efficiency of your sentences and it'll be as focused as possible on exactly what you wanted your explanation to achieve. There is, though, one more aspect of the tightening process.

GET A SECOND OPINION

As my long-suffering friends, family and colleagues know, I'm very keen on second opinions on any type of explanation. It's not always possible, depending on what you're preparing, but if it is possible, show someone what you've got. I do this for a number of reasons.

In the same way that 'tightening' the script digs out all the moments when you're not being as precise or efficient as you could be, getting a second opinion can often add an extra level of nuance and clarity to what you're saying. Even if this is a subject you know well, it's always worth checking that your explanation is accurate, fair and comprehensive. You also want to check that you're not assuming too much knowledge. If you are an expert in a particular subject, it's easy to forget what the uninitiated don't know.

When I send scripts to colleagues, I often get replies which start with, 'This is a small thing but . . .' To which I always reply saying I want to hear the small things! They're the ones that really raise the standard. We've talked a lot about how your credibility connects to the level of engagement your explanation will receive. Getting the finer points absolutely correct drives credibility. Equally, if you miss one error or there's one line of explanation that doesn't add up, it can undermine the credibility of the *whole* effort. You'll have experienced how this works yourself.

We've all had moments where there is a news story about something we know particularly well – perhaps it's about our home town or the industry we work in. If that story has an error in it, it undercuts your faith in that reporter's ability to get all the other details right.

A second set of eyes and ears is also a great test of the story you're trying to tell. This is, after all, a story that *you've* constructed and so you'll find it easier to follow. What you thought made perfect sense may not do to them.

'What I'm trying to get across is X,' I'll plead with Michael Cox and Mary Fuller with whom I make our explainer videos. I'll be trying to keep one line or another that I'm holding dear.

'I'm afraid that doesn't come across at all,' Michael and Mary will tell me. And it has to go – or at the very least, it has to change.

Lastly, second opinions can be a confidence boost. Let's imagine we're preparing a presentation in an important interview. That presentation doesn't just need to be a great distillation of the ideas and narrative that you want to communicate – it also needs to be delivered with confidence. Someone whose opinion you rate endorsing what you're about to do can have a marked impact on your delivery.

This is certainly true of my explainer videos. As I've experimented with 'assertive impartiality', the jeopardy for me and the BBC has increased. If you're going to be more assertive, you'd better get it right. It's not only about being factually correct and impartial (both of which we take great care over), it's about whether I'm *appearing* to be partisan in any way. These are sometimes fine judgements and I'll deliver the whole thing with greater conviction if I have a valued second opinion that says I've called this right.

I should add, I don't just do this at work. If I have a difficult email to send or a difficult conversation to have, I'll ask my wife, Sara, if how I'm planning to explain myself adds up. Her reassurance means I've greater confidence in what I'm about to say.

If you don't have people to whom you regularly turn to share your endeavours, it's worth thinking about who they could be. A good place to start is asking, 'Who do I know who does it well?' And this can be give and take. I've lots of colleagues and friends who I ask for help and who lean on me too. If you're really interested in explanation, it can be fun to exchange notes on what's working (I know how to have a good time . . .). If you're still struggling to think who you can ask, take a further step back: who could you ask about the fact you're struggling to find someone to help you with this? The more you ask, the more likely you'll find someone.

QUICK CHECK

- Are you satisfied you've tightened this as far as it can go?
- Is there anyone you'd like to show it to?

Your explanation is now going to be in very good shape. You could definitely use it. There is, though, one further step – because while explanation is about substance it's also about style.

STEP SIX

STEP SEVEN: DELIVERY

Human communication is a magical mix of art and science and if I were to focus only on those criteria that I set out at the start – simplicity, detail, efficiency, precision, explaining each detail, purpose, addressing the complexities – I'd have an accurate, to-the-point explanation. The question is: would people want to consume it? Would they *choose* to consume it?

How do we avoid making people feel rushed or overloaded? Does an 'efficient' explanation simply become too much? In the case of my early efforts on British Airways in-flight radio, this was certainly the case.

Do we risk offering an explanation that may be a decent distillation of what we want to say, but is the explanatory version of some undercooked greens – good for you but not much fun to eat?

This risk is the reason that 'engagement' is included in my anatomy of a good explanation. It's also the reason that I never assume I'm going to successfully explain myself. Not at least until I've been through Step Seven. My time at university and my early forays into journalism taught me some useful techniques to sift and collate information. What took my ability to explain further was when I started to understand that *how* to *keep someone's attention* was just as important as the information I was preparing for them. We've already talked about how the 'dial test' makes sure we're thinking about this and how 'joining phrases' can avoid those dreaded 'hard stops'. But there's something more we're reaching for here – and, again, it was music and musicians who helped me see it.

Not for the first time, let's turn to Los Angeles in the 1970s. We've already been there to think about how Joni Mitchell's music communicated with us on the album *Blue*. This time, I want to talk about one of the great electric guitar players you may never have heard of.

In the seventies in LA, top session musicians were commanding such good rates that they often opted out of being in one band. Instead, they worked with whoever they chose. It gave them a broader musical experience and, sometimes, it paid better too. Dean Parks was one of these session musicians and, through his career, he's been called up by the biggest names.

He played on Michael Jackson's album *Thriller*. He's performed with Elton John, Stevie Wonder, Madonna, Billy Joel and many others. It's a seriously impressive list. And when the group Steely Dan set about making what would become their 1977 album, *Aja*, they also turned to Dean Parks.

The context here is that Walter Becker and Donald Fagen, who made up Steely Dan, were infamous for burning through session musicians in their search for the sound they wanted on each track. To say they were sticklers for detail rather undersells the lengths they would go to. Sometimes a long line of session musicians would take their turn just to settle one solo on one track. Becker and Fagen were also trying to make music that was played on the radio and that sold well. This wasn't bubblegum pop music but equally it wasn't an exercise in deliberate obscurity. By the time they were making *Aja*, Steely Dan had already released five albums which had reached the US Top 40. No doubt, they hoped this new effort would follow suit – and it duly did.

I love this album and so needed no persuading to watch the *Classic Albums* documentary about how it was made. Walter Becker and Donald Fagen sit at the mixing desk with all the original elements of the tracks and talk through what they did back then – intercut with some of the musicians they worked with, including Dean Parks.

I'm reasonably confident he wouldn't have imagined that the interview he was giving about 1977 would feature in a book about explanation many years later, but that's what's about to happen.

We see Dean Parks sitting in a suitably LA scene – by a pool in the sunshine – as he explains the working experience of playing with Steely Dan.

> One interesting thing about Donald and Walter is that per-fection is not what they're after. They're after something that you want to listen to over and over again. So we would work then past the perfection point until it became natural. Until it sounded almost improvised. So it was a two-step process. One was to get to perfection. The other is to get beyond it and to loosen it up a little bit. It's quite an amal-gamation, that's for sure. And it's interesting to know that it can be a hit.

This was another of these moments that hit me straight away. I rewound the documentary and watched it again. I'm slightly embarrassed to admit, I felt a surge of excitement and my eyes widened. I was launching my programme *Outside Source* at the time and I had sometimes struggled to explain exactly what I was hoping to achieve in terms of style and tone. On the one side, I wanted the programme to be higher protein, more effi-cient and more detailed than some TV news. On the other, I wanted it to feel more informal, more approachable and more spontaneous than some TV news. On the face of it, these two ambitions were in competition. I knew that they weren't, but I'd been struggling to express it. Dean Parks captured this much better than I was managing.

The two-stage process that Dean Parks describes is exactly how I approach any form of explanation. The first stage is Steps One to Six. Getting everything as tight as we can. But, as Dean

Parks puts it, Steely Dan were trying to create something that people would 'want to listen to over and over again'. Now, even my mum doesn't watch my explainer videos or TV programmes over and over again. My goal is to get people to really listen to us once. But the point still stands. To get people to listen, we need the second part of the process – to, first, get everything perfectly lined up, and then to 'loosen it up a bit'. That is Step Seven.

When *Rolling Stone* reviewed *Aja*, it described how Steely Dan had shifted from 'the pretext of rock and roll toward a smoother, awesomely clean and calculated mutation of various rock, pop and jazz idioms'. Clean and calculated. This, again, resonated hard. But as Dean Parks noted about the album, 'it's interesting to know that it can be a hit.' It was – selling more than any of their other albums. Steely Dan had mastered being both tight and loose.

All of this may feel like a sizeable and slightly random tangent away from the nuts and bolts of constructing an explanation. In fact, it's right at the heart of it.

As my explainer videos were first becoming popular, the Reuters Institute for the Study of Journalism invited me to do a seminar about them. The associate director for its fellowship programme, Caitlin Mercer, was the host.

'There's something I noticed while watching your explainers,' Caitlin said to me as we got going. 'I noticed I sort of started keeping time and I noticed at around every eight seconds, something new is happening: so a graphic is changing, a new voice is coming in. And it's almost like a song. Is that conscious?'

The answer was a resounding 'Yes'. I am acutely aware of the rhythm of how I deliver explanations. The best written and verbal explanations have a rhythm, a personality and a fluidity that complement the precision and relevance of the information they are passing on. Here's how we can try to make sure your explanation has that.

VERBALISATION

One of the best ways to check our explanations is to use the knowledge that we all have about how we communicate with each other. Earlier, I was saying that I always ask of all my explanations: 'Would I talk like this?' And, if the answer is no, it has to change.

There's something similar with how explanations flow. This can be hard to define but a story, essay, speech or argument should have a 'flow'. The vast majority of the communication we receive is spoken and we're all expert at it. We know when someone is making sense to us and we can also tell if someone isn't. When there's a 'flow', we'll find it easy to follow – there will be a natural progression through the information we're hearing. If we stop to think about it, we can spot when it's there and when it's not. Even if we're not sure why, it's possible to feel it.

As you'll have already noticed, I have a preoccupation with how we join information together. When I use the word 'flow', that is really what I'm talking about – the way that we take who we're talking to through the explanation. When the flow is right, there is a logic and rhythm to your words. If the flow is off, the joins will almost judder – either because they don't quite make sense or because the rhythm of the words doesn't match what's gone before and what comes after.

This may sound like an elusive concept but it's not as elusive as you might imagine – because you know when you speak clearly and you know when someone is making sense to you. Here we can use those instincts. We do so using verbalisation.

I verbalise every explanation that I write, whether it's to be read or heard. For me, however hard I concentrate when reading an explanation in my head, it doesn't come close to being as effective as saying it out loud. Even after twenty years as a news presenter, I still cannot be sure a script is quite right until

I've said it out loud, often several times. When I do this, I almost always make changes.

I think of this as ironing out all the creases. It's not about the big ideas in our explanation or the facts or the structure – we're happy with all of those. This is about the flow and how well the whole thing hangs together. This isn't only about improving the way the information is delivered. In the case of spoken explanation, it will also help identify stumbles or sections where you lack precision and punch.

Start by opening up your explanation and reading it at a slow-to-medium pace.

Don't rush through. You want to have the chance of picking up every crease. As you read, ask yourself:

- **Does this sound like me?**
  If it doesn't, is there an adjustment you can make so that the words feel like your own?

- **Does each sentence logically move on to the next?**
  If they don't, try adjusting the end of the first or, often more easily, the start of the next one to make the join work better.

- **As each sentence follows the one before, does it feel right to you?**
  If it doesn't, can you pinpoint the word(s) where the flow doesn't feel right? What is it about how those words sound that isn't quite right?

After each adjustment, read it out loud again. If it's worked, great. If it hasn't, keep trying until it's right.

There is one simple rule of thumb in this part of the process. If it's not right, don't leave it. These 'creases' can really affect the way you deliver explanations. Remove them all and the overall effect can be very powerful – both for how comfortable it is for you to deliver and for how clear and confident you will sound.

FROM TOP TO BOTTOM AGAIN . . . AND AGAIN

'Another time?' you ask. I'm afraid so. I think of this as the bike-repair-shop moment. You know, when you take your bike for a once-over and the mechanic checks the chain, the cogs and so on. Once their work is done, they raise it on to a stand and spin the wheels and, all being well, it whirrs smoothly.

Read your explanation out loud once more. All being well, your explanation is going to be smooth too. If it is, nothing about your explanation is going to change (which is a very good feeling). If it isn't, what's causing the crease? Normally you can pinpoint a word or two. Alter them and read it out once more.

If yours is a written explanation, your work is done. It's complete. It's ready to use. That's a feeling I never get bored of.

If yours is a verbal explanation, you're almost there but not quite. First, let's look at the visual elements which will accompany your speech or presentation, if you're using them. If you're not, you can head straight to page 188.

PLACING YOUR VISUAL ELEMENTS

In Step Four, you made an initial list of visual elements that you thought you would add to your explanation. Now it's time to revisit them. We need to decide which of those elements we want and where they are going to go.

Open up your script and your list of visual elements and start marking on the script where they should go.

As you do, keep in mind two things.

First, consider what is behind you at all times. If you bring up, say, an image that supports your words and then move on in the script, that image will still be there but will no longer match what you're saying. Some visual elements will work across a longer passage of your talk, some will only work for the moment you're explicitly referring to them. I'm always on the lookout for those visual elements that can be a bridge through sections of my explanation.

Second, timing is crucial. I'm a stickler for this. If you can get your timing precise on visual elements, then they will have far greater impact.

This is something I spent a lot of time working on when we created *Outside Source*. As I've mentioned earlier, the idea of the show is that we collate all the most essential information on a story in whatever form it comes. It's then my job to weave it all together. The challenge is that there are often many elements – just as there may be in your explanation – and if they are visible at the wrong time, it's both distracting and incoherent.

For many years, we had a touchscreen which I'd use to pull up the elements. To make sure the timing was just right, I put instructions in the script. I use something very similar for any explanation I'm giving that involves visual elements. Here's a simple fictional example of how it works with the touchscreen.

EXAMPLE

INTRO
A large protest has begun in Paris.

MAP
It's taking place around the Arc de Triomphe which you can see on the map here. And there have been clashes.

IMAGE
This is one picture taken by a Reuters photographer. Here we see regular police.

IMAGE
But in this second picture we can see riot police too. They've been sent in by the City's Mayor

QUOTE
. . . who has called the protestors 'trouble-makers'.

VIDEO
And look at this – this is a video from a side street where more and more people are trying to join the protest.

You can see how I would position the instructions exactly when I wanted to introduce the elements – not necessarily at the end of the sentences. There is a direct connection between my words and what you would see. The combination gives what you're saying the greatest chance of impact because the words and visuals are working together for a common purpose. You can also build momentum in your story.

If you were telling the story of how you built your company from scratch and wanted to show its profits in 2010, 2015 and 2020 – if you placed all that information behind you at once much of the drama is lost. Instead you could do this:

EXAMPLE

GRAPHIC SHOWING THE DATE
I created the company in December 2008. At first it remained small.

GRAPHIC OF 2010 PROFITS
In 2010, our profits were £50,000. We reinvested that.

And borrowed more. By 2015, though, we'd expanded across the EU. Our profits were . . .

GRAPHIC OF 2015 PROFITS

. . . £1.5 million. We hired more staff, paid off our debts . . .

PICTURE OF ORIGINAL HQ

. . . and moved from our tiny offices in West London . . .

PICTURE OF NEW HQ

. . . to much bigger offices down the road. And by 2020, our profits . . .

GRAPHIC OF 2020 PROFIT

. . . were £5 million. We were bigger than I'd ever imagined we could be.

Each element builds on the next – supporting the information you want to communicate and building a narrative that, we hope, the audience wants to hear more of.

Doing it this way, you end up with lots of visual elements and lots of slides. That's fine. It doesn't mean you'll take any longer; it just means that your visuals are much more finely tuned to what you're saying.

Having chosen your elements and ordered them, try talking them through with the script. Can you feel that the two are in sync? If not, adjust the visual elements until you feel they are always supporting what you want to say. Keep trying until you're happy both with their position and your ability to bring them up as you deliver the words.

I use exactly this technique in any presentation I'm giving.

One final thought on this – if your words and the elements are relevant and riveting, you don't need any bells and whistles. Don't use any fancy sounds or animations unless they actively add something to your explanation. Otherwise they risk distracting and sending a message that you need them to make it interesting. If what you have to say and show is relevant, clear and efficient, more likely than not, your audience will pay attention.

By this point, you are a long way into preparing the delivery of this explanation. You've read your work multiple times. You've checked its flow. You've removed any creases. You've also chosen, ordered and timed any visual elements you may need.

And these visual elements are part of your next calculation – how you plan to make sure you say exactly what you want to say.

SCRIPT v BULLET POINTS v MEMORISATION

Staying on script

At the moment, you've written your explanation out in full. As I mentioned earlier, there will be some circumstances – a set-piece speech, a formal work presentation – where you want to have every single word nailed down. In which case, what you have doesn't need changing. But do think carefully about whether this is the best option. However well you read those words, unless you have some form of autocue (or teleprompter as they're also called) you will struggle to disguise that you are reading.

(Barack Obama always used a teleprompter for speeches, as do many US politicians. They use the transparent ones that are set just to the right and left of the lectern. That's why if you watch their speeches, they look to one side and then the other. It allows them to deliver a speech in a way that is both natural and on script. Unfortunately for most of us, this technology is unlikely to be available when we give a talk.)

Reading word-for-word has its attractions. You can guarantee the precision and clarity of your language. You can time your explanation very well. You have complete control. But there are risks and downsides you should be aware of.

The first is a practical one. When you look down at the page, a lot of words will greet your eyes. In that split-second, you need to be able to find your place to continue. This is hard when you're practising. It's harder if the pressure is on. Even as I've become more confident with public speaking, I know well that feeling of looking down and not finding my place instantly. It can take less than a second for the mind to start racing. And while your eyes scan frantically for the place you want to pick up from, you need to pause, which in turn can impact the rhythm and delivery of the explanation.

If you do need to read a script, here are three practical measures that can help. Ask yourself:

- **Is the font size big enough to make the words easily discoverable at a glance?**
  Remember, you're not laying this out for anyone other than you to read. It might be a little ridiculous to publish an article or produce a handout or pamphlet with twenty-point font. That isn't what you're doing here. The words only serve one purpose – to help you speak. Making them big enough to read easily is helpful. Though the bigger you go, the more page-turning you will have. You'll soon enough get a feel for what works for you. And you won't be the only one doing this. If you ever see major speeches being given when the speaker is reading off paper, they tend to arrive with a thick wad of A4. Even experienced operators prefer bigger fonts for these moments.

- **Are the lines spaced out?**

  Lines with minimal spacing between them (such as this book and, indeed, most books) make it harder to locate specific points in the text. Not a problem if you're reading a book, but it can be a problem if you're explaining yourself from a script. Again, you'll work out what feels best but I prefer 1.5 line spacing if I'm reading direct from a script.

- **Have I added headers?**

  Headers are very common in many forms of written text from this book to an article in a magazine or on a website. In those cases they're used to signal the subject of the section that you're about to read. In this, they are there only for you. They are very helpful if you need to find your place in an instant. Add as many as you find helpful. I tend to not hold back.

Having a fully scripted explanation will feel right sometimes. For me, though, and I suspect for you, a lot of the time using a word-for-word script is not going to create the effect that you're looking for. It risks being overly formal and constraining your ability to connect with the people you're addressing. It's also not the easiest way of accessing the explanation you've prepared. That's where bullet points come in.

Bullet points

Whenever I prepare to give any form of verbal explanation, if I have notes or a script, I always check if the words I'll be looking at are there to make me feel better or to *actually* make me better – because these two things are not the same!

Word-for-word scripts can feel comforting but a lot of the time they won't make your delivery better. Bullet points can do that. Having only the most important words in front of us is, in many cases, more likely to achieve that.

If you do want to prepare bullet points, copy and paste a new version of your full explanation. (We want to keep the full script as a reference, so don't delete it.)

Now take each section in turn.

How much can you strip out and still remember what to say?

Which trigger words will be enough for you to remember that section?

If you've stripped too much out or don't have enough trigger words, put some back in until it feels right.

Experiment with what you have and, remember, the first time you try this is very unlikely to be your best. I often start with more bullet points than I end up with. As I practise and it becomes more familiar, I strip the bullet points back.

At first don't try to do the whole explanation. Initially, practise each section on its own – and allow yourself to look at the bullet points all the time. When you feel comfortable with each section, then practise stitching two together. When that feels comfortable, add more sections and so on.

Keep adjusting the bullet points as you do this. They have no other purpose than to help you deliver a fluent and precise explanation. When you feel you're approaching that point, go through the process again but this time imagining you are addressing your audience. You will now be looking away from the bullets frequently. This, of course, is much harder. But you might get a nice surprise. By forcing yourself to talk through the sections with no script and only bullets, your brain will have become much more comfortable with constructing these sentences in the moment than if you'd been practising with a script.

The sentences you're saying out loud will, if you've written this explanation out in full, be based on sentences that exist. As such, we want them to be similar to what you've already decided to say. Each time, though, they will vary a little as you deliver them. That's not a problem. In fact, this is where something magical can happen.

The key phrases of the explanation that you've produced will start to become incredibly familiar. You will always say them almost exactly the same way. But because you're using bullets and not a script, the words that you place around these key phrases will inevitably vary and will feel natural and spoken (rather than scripted and read). When this works well, you land the lines that matter most with confidence, clarity, control and consistency, but in a way that is far more engaging because you're speaking to people not reading to them.

As you practise, you'll also start to notice the moments when it is most natural to glance down. You might want to note those or at least make a mental note of them. Glancing down is both a moment to get help but it's also a moment of danger – when your explanation could lose purpose and engagement. The more you know about when and how you're going to look down, the more the bullet points can help you.

As you can probably tell, I'm an enthusiast of bullet points. I use them all the time from major set-piece presentations to live reporting on TV and radio to a talk where I've had minimal time to prepare. They are a really effective way of achieving both precision and fluency.

Here's an example of how I'd prepare bullet points. This is a section of a talk at a media conference that I gave in 2022 not long after Boris Johnson announced he'd stand down as Prime Minister. For various reasons, I decided to do this one from a full script. But had it been a more informal setting, I'd definitely have used bullet points, and this is how I'd have prepared them.

EXAMPLE

What I wanted to say

Delighted to be here to talk to you about the future – and about news. I want to start by going back to Wednesday 6 July. It was a warm summer's evening – and I was standing in Downing Street.

A few metres away was the famous black door of Number 10 – and behind the door was Boris Johnson. The PM was in deep political trouble. Many of his colleagues wanted him to go – and in that moment, we all wanted to know one thing – would he go?

But of course I didn't know. I certainly didn't know that the next morning Mr Johnson would bow to the pressure.

Historians may decide his departure was inevitable. But standing in Downing Street – staring at the windows with the lights still on that dramatic Wednesday night, I didn't know.

But I did know plenty of other things. I knew what Conservatives were saying. I knew what the best Westminster journalists were reporting. I knew the processes for Boris Johnson to survive and for him to go. I knew the long-term reasons why he was in trouble. Whatever the next development, whatever the future, I was prepared with the information that gave me the best chance to respond to the story.

First set of bullet points

Here I have done two things. I've taken out words that aren't needed for me to get the gist of the sentence. I've also taken out information I'm certain I'm going to remember. For example, I am delighted to be giving the talk, but I know I'm going to remember to say that!

Intro

- Here to talk to you about the future – and about news.
- Wednesday 6 July. Warm summer's evening. Downing Street.
- Famous black door of No. 10. Boris Johnson in political trouble.
- Many of his colleagues wanted him to go. Would he go?

My situation

- Didn't know that the next morning Mr Johnson would go.
- Historians, inevitable.
- Downing Street, lights, Wednesday night.

What I did know

- What Conservatives were saying.
- What Westminster journalists were saying
- The ways that Boris Johnson could survive or could go.
- Long-term reasons why he was in trouble.
- I was prepared with the information that gave me the best chance to respond.

Stripped-back bullet points

By this point, I'll have rehearsed the speech several times. The phrases, descriptions and points I want to make are very familiar. So all I'm looking for from the bullet points are triggers. These will be enough to prompt me to say whole sections that are very familiar.

Intro

- Wednesday 6 July. Downing Street.
- Would he go?

Situation
- Gone the next morning.
- Inevitable?

What I did know
- Conservatives
- Journalists
- Process
- Long-term
- Best chance

Or if I was feeling really good . . .
- Wed 6 July
- Situation
- What I knew

Those three reminders may be enough to trigger everything I want to say in this section of the speech.

ASK YOURSELF
- Is the font big enough?
- Is the spacing right?
- Can I easily see bullets at a quick glance?
- Can I talk through each section fluently with only the bullets for prompts?
- Can I talk through the entire explanation fluently?
- Would it be better if I added or deleted bullet points?
- Am I confident when I want to look down?

Memorising

As well as a script and bullet points, there is a third option available to you – to memorise the whole thing. Now, if you're giving a lengthy talk with no visual accompaniment, this is quite an undertaking. There are two ways to do this. You could memorise an entire script (in a way that an actor learns a script). I've never done this and, while I am happy to be corrected, can't think of any situation when you would need to be word-for-word perfect. Let's leave that to one side. The second version is essentially to prepare bullet points as we did in the previous section and then to learn both the bullets and what you want to say around them.

It's not impossible and can lead to very convincing explanations – but always ask yourself, why do I need to memorise all this?

If there's not a major need, I'd make things easy for yourself and stick to written bullet points.

If you've got a good reason, we'll look at memory techniques that can really help take this on a little later.

Before that, though, there is one technique that allows you to speak without using any form of notes but doesn't require every last detail to be committed to memory.

VISUAL PROMPTS

The difficulty level of talking from memory reduces sharply with a visual presentation to accompany you. In many cases, the slides (or any other visual support you're using) can act as a prompt which makes being completely without notes much easier. They act almost as a replacement for the bullet points that you might have had. That's the good news. There is a danger here, though.

The most convincing explanations have purpose and momentum. An explanation that pauses for a prompt every few seconds will not have either of those. I'm sure you can think of PowerPoint presentations where the speaker talks to a slide, pauses, looks at the screen, clicks on to the next slide, pauses again and then starts speaking again. It tends to not be the most arresting way of explaining anything.

As such, when we're talking to slides, we want to talk *into* those slides. We need to know what is coming. So while slides or any visual accompaniment can be helpful in speaking from memory, we still need to know what we want to say. We can't rely on the slides to tell us what to say next. I'd go further than that. That is not the role of the slides. Any visual accompaniment may well make your life easier by helping trigger what you want to say – but the reason it's there is merely to support and add to what you're saying, not to become the driving force of the explanation.

If you choose this route, practise talking through the slides. As you do, the order will become familiar and you'll be able to talk into them rather than be prompted by them.

Now that you've decided whether you want to use a script, bullet points or to do it from memory, you're on to the final piece of the jigsaw – *how* you're going to talk through your explanation.

PACE

There are a number of ways to calibrate how you deliver an explanation. Let's start with pace. This is yet another lesson that I learned the hard way. In my early days as a radio presenter, in my eagerness to be engaging, I sped up. Speed was also seductive, as the faster I spoke the more information I could include.

Both calculations were wrong and, not for the first time, the audience may have wondered what had hit them.

To try to address this, I started to experiment with using the same engaged tone but lowering the speed. I still sounded actively interested in the subject – but the audience had more space to take in the information. My presentation improved.

I also traded speed for a focus on efficiency. If my language was efficient, then I would create more time for information even if I lowered the speed of my delivery. This helped too.

There was an added benefit. As my delivery slowed, my breathing improved, and if your breathing is controlled, you'll speak with greater authority.

It's worth experimenting with the pace and tone that suits you. The two best ways of testing how you're doing are to record yourself and listen back or to ask someone to listen to you. If you feel shy about either, I can relate to that, but the rewards will outweigh any awkwardness you may feel.

If you do record yourself, my advice here is to make two columns – one for what you'd like to keep, one for what you'd like to change. Self-assessment can't just be making a list of everything you're getting wrong. You'll be getting things right as well and it's just as important you note those so that you can keep doing them.

Once you've settled on the pace and tone that you're happy with, experimenting with shifts in pace can be really effective too.

EMPHASIS TECHNIQUES

Explanations benefit from emphasis. Some parts of what you have to say are more important than others. Pace, pauses and intonation can help mark that.

Here is an example:

*This fall in sales was predicted. The industry body released this statement saying: 'We're very disappointed with what has been allowed to happen but sadly this was the only possible outcome once the regulations were brought in. We will be taking **legal action**.' And look at the numbers it released. Sales in the sector were worth **£100 million two years ago. Last year they were £10 million.***

The first part of the quote is relevant but relatively predictable. I'd speed up through this. The three pieces of information in bold are the most important – the threat of legal action and the stark comparison. I'd look to slow around all three. Here's the text again, but I've marked up how I'd deliver this – with bold for emphasis.

[Start at regular speed] *This fall in sales **was** predicted. The industry body released this statement saying:* [minor pause, speed up] *'We're very disappointed with what has been allowed to happen but sadly this was the only possible outcome once the regulations were brought in.' It goes on:* [back to regular speed] *We will be taking **legal action**.'* [Minor pause] *And look at the numbers it released. Sales in the sector were worth* [slower] *£100 million two years ago. Last year they were* [minor pause] *£10 million.*

If you read that through following the instructions, hopefully you can hear how the shift in pace and emphasis can help communicate what is more important.

Needless to say, I'm not expecting you to go through every explanation you give plotting every point of emphasis in this detail. But it's worth thinking about what matters most and how to emphasise that in your delivery. As with many of these techniques, experiment with what works for you.

SCRIPT-MARKING TECHNIQUES

When I joined the BBC in the 2000s, radio newsreaders would all mark up scripts to help them deliver the news summaries. I'd watch, fascinated, as the newsreaders would come into the studio for their bulletins. More often than not, across the desk I couldn't make out the words of their scripts, but I could see the marks they'd made around the words. I'd never seen this before, but I had reason to be interested.

Whenever any of us are nervous, our breathing is likely to become shallow. If that happens, we need to take more breaths than we normally would. In turn, if that is happening, we're more likely to do it right in the middle of a sentence. All of which can really disrupt how you deliver your words.

We'll discuss how to manage nerves a little later on, but whether you're nervous or not, planning how you are going to deliver your explanation is going to give you much more control and confidence in how you do deliver it – and breathing is a part of that. Your use of short, sparse sentences earlier in the process will already help but I use script-marking to give me an extra hand as well. Here's how.

/ Pause

I don't mark every pause that I'd naturally take between sentences. However, sometimes if we don't pause at a certain point, we may run out of steam before we next have a chance to. I put forward slashes to tell me to breathe.

The house was uninhabitable. It was damp, the windows were smashed, the roof was leaking and, frankly, it stank. / We had to take the decision to knock it down and to start all over again.

// Pause for emphasis

I use double forward slashes to prompt a pause for emphasis. It's longer than you might normally leave at the end of a sentence.

We had to ask ourselves several questions: //
Did we have the money to do this? //
Did we have the contacts to do this? //
And did we have the stomach to do this? //

→ Keep talking

This is a really important one. Often our flow, rhythm and breathing can be disrupted by pausing at the wrong point. If there's a place where it's crucial that you keep going, then draw an arrow underneath at the point where you might be tempted to stop. If you don't plough on, you may pause briefly and then not reach the place where you can take a proper breath.

The area is beautiful. But there were three issues: the river was polluted, the fields were polluted and the air was polluted.

Underline for emphasis

If there are words that require emphasis, make them easy to spot.

The costs of the project had increased by 50 per cent.

I'm not suggesting you mark up every word of every verbal explanation you ever give. There will be times you'll feel confident enough without this. There will be times when it's not realistic to be looking at your script long enough to see all the marks. (Both newsreaders and presenters on the radio can look down the whole time and so have the huge advantage of not needing to worry about looking at the people we're addressing.)

Some of these rhythms will also become natural as you become more accomplished as a speaker.

However, marking up is a really useful way of thinking about how you want to deliver your words. I'll often mark up a difficult or important section of an explanation in advance so I can practise it. I'll experiment with pace, pauses and emphasis and, once happy, I'll mark it and go over it a few times. Once I'm well grooved in how I want to deliver it, I might not actually use a marked-up script in the moment. The shape of it will now be so familiar it'll happen without the need for marks.

I should add, I made these marks up and you should feel free to do the same. Their only purpose is to help your delivery and you'll know what works best for you.

By this point, you're really starting to know your explanation well. There's one more thing to keep an eye out for.

TRIPPING UP

You can have done everything right to this point and when you say a certain phrase, name or point, it doesn't come out right. It's just not rolling off the tongue and doesn't feel comfortable. This matters because there's a good chance you'll trip up – or at the very least be preoccupied that you might, which, in turn, could affect how you deliver this section.

There's even a well-known phenomenon that TV presenters are familiar with where you will prepare to pronounce a difficult name correctly, manage to get it right, but stumble on the words that immediately follow. Your brain has so focused on the problem, it doesn't get the words on either side quite right.

If I think back to my teens, one of the things my piano teacher always told me when I was struggling with a section of a piece was to slow it down and go over it until I got it right. This is sound advice – and, if the potential stumbling block is un-avoidable, this should be your first move here too. See if marking up the script helps you navigate the sentence or two where the problem is. If it's a difficult pronunciation, write it out as it sounds and say it slowly over and over again.

However, there is a difference when learning a piece of music. If you were, say, learning a piece on the piano and there was a section you kept getting wrong, you would slow down and practise each part of that section in isolation, but you couldn't actually *change* the notes. That option isn't available to you. Happily, for us here, it very much is. There's very rarely one way of saying something and, so, if your mind is aware of an issue with a phrase or name, change it if at all possible. Move away from it.

When I was getting going as a presenter, Mahmoud Ahmadinejad was the President of Iran and, however well I could say his name in the office, when I was in the radio studio I would stumble. Every time. Either on the name, or just before or after. It became a thing in my head, which wasn't ideal as he was in the news a lot. With each stumble a colleague would, in a friendly way, send a guide on how to say the name. But that wasn't the issue – I could say it perfectly outside of the studio. I decided to move away from the problem.

Each time, the name came up in a script or an interview, I'd reference the Iranian President or Iranian leader but not say his name. As far as I know, no one noticed, and I presented all the stories about Iran with more confidence. Within a couple of months, the issue had lost its sting and I started to drop in the occasional 'Mahmoud Ahmadinejad'. With each one my confidence built, and the issue was soon gone. That was helpful for a time, though, once he was out of power, my ability to say his name with confidence wasn't quite as useful as it might have been.

Never feel bad about simply removing a potential stumbling block. An audience only knows what you say, not what you were planning to say. So long as you've not compromised the essential information you need to get across, moving away from the issue is often worth doing.

This is one way of making yourself feel comfortable with how you're going to deliver your explanation. In their different ways, everything in the 'delivery' section of this process is designed to do that. The connection with familiarity and the quality of your delivery is direct. This applies to the words you're planning to say – and where you're going to say them.

But before we get to that, a word about time-keeping.

STICKING TO TIME

It's hard to overstate the importance of this one. If you're sure what you have to say fits into the time you have, that's one more thing off your mind.

You also reduce the risk of having to rush or jettison important information to avoid running over. All this will boost the confidence and conviction of your delivery. And it means you say *all* of what you planned to say.

The flip side of this is that rushed delivery, last-minute ditching of elements or simply crashing through the end of your allotted time does nothing for the confidence the audience has in you.

Now, the trick here is to make sure you've accurately timed what you have to say. That's harder than you might think. This advice comes from experience.

Sometimes I'll be asked to estimate the timing of a script for a report we're making. I'll sit at my desk in the newsroom and read it through with a stopwatch going. 'Now can you do it as you're actually going to say it?' one of the producers will ask with a twinkle in their eye. They know I've got form for reading slower when it's the real thing. Which means I risk telling them the report will be, say, three minutes, when in fact it will end up being four. Being realistic with your estimates is a valuable discipline.

I'm normally very tuned in to getting this right when public speaking. I do lots of it and I very rarely bust the clock. But the other day I was reminded once more of the perils of getting your timing estimate wrong.

I was giving a speech at a conference in Copenhagen. I'd done everything that I'm advocating in this book in terms of preparation – and when I timed myself in advance it matched the speaking slot I'd been given.

When it was my turn to speak, everything seemed to be going well. I was in full swing when the moderator politely cut across me: 'I think we'll need to move to your final thoughts.' I'd fallen foul of two common traps. My delivery was slower than when, relaxed in my hotel room, I'd rattled through the timing of the speech. Second, sometimes we'll add extra words and phrases as we go through a talk. It can sound fluent and infor-mal, but it adds time (and makes us less precise). I'd done both of these things and, now, with a panel waiting to start, was out of time and having to ditch the conclusion of my speech.

The frustrating thing was that I could easily have cut it back in advance – but I hadn't spotted that I needed to. It was a sharp reminder to follow each step carefully and, in particu-lar, to always properly assess the duration of what you plan to say – and, if possible, to leave some leeway. This is especially important when giving presentations in interviews or pitches where you want to be in control and to get all your points across. Now every time I assess my timings, I think of Copenhagen . . .

FEELING COMFORTABLE IN YOUR SURROUNDINGS

When I arrived at university, I religiously listened to a BBC 5 Live programme called *Up All Night*. I mentioned it earlier when discussing authentic presentation. *Up All Night* covered an eclectic array of stories from around the world and, whatever I'd been up to in the evening, I'd switch it on and be taken here, there and everywhere as I fell asleep. I loved it. It made me want to work in radio and it deepened my desire to be a global news journalist. Ten years later, to my delight, I was a producer on *Up All Night* and, having already contributed segments to the show, I was asked if I'd cover for one of the main presenters one weekend. Of course, I said yes and then had several weeks to ponder what was coming my way.

My first go at *Up All Night* was, I think, a Friday night. This wasn't just my first turn on this show, it was my first time presenting anything on the BBC. Not just that, it was 5 Live, a network I had gorged on for years – not least when I'd been unemployed a few years earlier. By any measure, this was a big deal for me and, as the moment approached, I felt anything other than OK. I was preoccupied, felt profoundly out of control and was, to use the non-technical term, in a huge tizz.

'What can you do to help this?' my wife, Sara, asked. As I replied, it was clear I wasn't nervous about doing the interviews – I'd done quite a few of those on air before. What I was anxious about was everything that came with being at the helm of the show – how to start it, where the music came in, when I went to the news and so on. 'The furniture' is the jargon we use to describe the fixed elements of a programme. I knew the furniture as a producer, but I didn't know how to present within it.

'I think you need to go in,' Sara suggested. 'Can you get access to a studio?'

I could. I arranged to book one and for all the music I might have to talk over to be available. Feeling slightly silly, on a day off I travelled an hour from our place to the BBC. When I got there, I sat in the studio and practised everything again and again. From the more complicated aspects, such as the start and end of the show, down to the more straightforward tasks, such as ending an interview and talking into the news. I even did the easier ones repeatedly until I was really comfortable with them. I also familiarised myself with where I'd put my notes, where the computer was, where the microphone was, where I'd be looking to see my colleagues through the glass in the gallery. A couple of hours later, I headed home.

That Friday, I topped that up. I made sure to be in the 5 Live studio as early as I could. Our very patient sound engineers helped me go through the furniture further times. And then the clock edged inexorably towards the start of the show.

Don't get me wrong, I still felt a long way out of my comfort zone. It *was* scary and my start wasn't perfect. But it was OK and I got through it because, though there were a lot of nerves, I was feeling nervous in surroundings that I was familiar with. That helped.

I remembered this experience ten years later, when I was asked to host a set-piece live TV debate for the 2013 German election. The whole set-up was intimidating. It was a big-budget production. There were, I think, five cameras, with one being on the end of a boom for the sweeping opening shot. We'd set up in the grand courtyard of the German Historical Museum in Berlin and had an audience of around a hundred coming. I'd rarely done a programme this high profile or expensive. Plus, it'd be live, so there would be no second chances. It's fair to say I was feeling the pressure. And in Berlin that week I applied the lesson I'd learned many years earlier that day at the 5 Live studio.

The day before that live debate, I decided to work in exactly the spot where we were broadcasting. As the crew assembled the set, I worked on my scripts right there (while being careful not to get in the way!). Then, when the set was complete, I talked to the producers about our plans right there on set. I began writing and rehearsing my scripts on set too. I wanted to try to feel at home there. Again, it helped when the moment came the next day. I can remember the director counting down 'three . . . two . . . one' and me being nervous but also in control.

I do versions of this all the time – whether broadcasting or otherwise. Even if I can't get into the space where I am going to be talking, I'll try to find out about it. I had to give a talk to the BBC's Executive Committee a few years ago. I'd never been in the room before and I wouldn't be able to see inside it on the day, as the Committee's members were in there behind closed doors. In the days before, though, I made a point of asking about the format and the layout, right down to where I'd be sitting in relation to the screen I'd be talking to.

My point here is not that you have to do this every time you do any form of public speaking. Not at all. I'm sure there will be times when you have no knowledge of what awaits and speak brilliantly. However, this is simply another thing you can do to make yourself comfortable. I find the more I do this, the better my chance of explaining myself well. It all adds up. What you wear is part of that too.

FEELING COMFORTABLE IN YOUR CLOTHES

Fear not. You're not about to be on the receiving end of styling advice. I know my limits, and this is a line I won't be crossing. But it's worth noting that feeling comfortable and confident in what you're wearing can influence how well you speak. Until relatively recently, I had minimal confidence when it came to selecting work clothes. I became aware that fretting about this was becoming a distraction. I'd be thinking about my clothes when I could have been preparing. Or I'd be fretting about them when I was actually speaking. Feeling comfortable in your clothes is, for me, as much about removing that distraction as it is about looking snappy.

I was so aware of the issue I sought out the help of a brilliant stylist called Jane Field (a decision which led to my ongoing commitment to dark blue). Now, I understand that a stylist may neither be possible nor necessary for you. I'm also sure you know a lot more about clothes than I do. My only advice here is to resolve what you want to wear in the different circumstances that you're likely to encounter. I have three or four outfits that I'm happy with for certain situations that keep coming round. It means that I don't waste time ahead of pressured moments making those decisions. If you don't know what clothes work best for you, consider who you know who could offer some advice.

If all this has gone well, as you prepare to deliver your explanation, you're comfortable with the environment in which you'll be giving it and your outfit is set. But there is still one small extra I always do.

HAVE A HANDS PLAN

People always laugh when I say this one but, believe me, a hands plan is worth having. Much like other advice I've given you, this is in part about what you can add to an explanation but, just as importantly, it's about removing distractions. And believe me, if you don't know where your hands are going to go, they may take on a life of their own.

A hands plan doesn't take very long.

The first question I ask myself is whether one hand or another is likely to go rogue. In my case, I'm right-handed and so I tend to use my right hand to gesture. If I'm sitting at a desk or standing at a lectern, I'd normally resolve to park my left hand on the desk and not move it. When possible, actually holding something like a table is a very good way of getting a hand to stay where it is. If I'm using a clicker in a presentation, I'd place it in my left hand, leave it down by my side at all times and use my right hand to gesture. If I'm standing in front of an assembly of children with nothing to hold on to, I might resolve to join my hands in front of me. I was interviewed for a job the other day and decided to place my hands in my lap and then use my right hand for emphasis if I needed to. When I sat down, it was helpful that I'd already decided what I was going to do.

None of this needs to feel too prescribed. What feels comfortable for me may be quite different to what feels comfortable to you. It also is not the biggest decision you'll take when preparing an explanation. Equally, we can all remember times when we're watching someone talk and we're thinking, 'Why do they keep doing *that* with their hands?' They can be a distraction.

We can also use hands to add punctuation. You may have three points you want to make. You could gesture to your right, in front of you and then to your left as you mark the points. You may want to emphasise certain words. If you were saying 'and all these factors', you could move your right hand from in front of your body outwards to the right to emphasise the expanse of factors. If you were saying 'but hold on, why did this matter so much at this point?', you could reach one arm out a little in the direction of the audience with your palm facing towards them.

You can try some out and see which feel comfortable. Avoid getting carried away, though. Less is often more with hands.

If the use of hands for emphasis becomes too obviously self-conscious it can feel inauthentic and become a distraction. But if used with restraint, it can be very effective.

Whatever you think works best, having a hands plan gives you the best chance of doing it in the moment.

HOW TO STAND

This is not going to be a long section, but it does warrant one. How you stand can have a huge effect on how comfortable you feel – and how those you're speaking to feel. Many of the most important aspects of communication are rooted in what we do in our normal interactions. The trick is to hold on to those instincts when we're in what feels like an abnormal setting.

In the case of standing, simply think about how you stand when you're talking to someone at a gathering or in the office. My guess is that your feet won't be pointing directly forward nor will they be perfectly parallel. We tend to only stand like that when we feel the need to be formal. And if your feet are pointing directly forward and are completely parallel, you're unlikely to feel as stable or as relaxed as you normally would. If you're not sure what I mean, quickly try it. Put your feet in parallel. See how that feels. Then just allow your feet to turn away a little to however you'd normally stand. My bet is that one foot is slightly in front of the other and that one is facing forward more than the other. Try to bend your knees a little as if you were checking your balance. That is going to feel a lot more solid than if you are more formally facing directly forwards.

The same is true of how we position our chests. When we talk to each other, we very rarely square up. We naturally place our bodies at a slight angle. No need to change a thing if you're

STEP SEVEN

doing any form of public speaking. Do what comes naturally – it will be better for you and your audience.

I use this all the time when reporting on location. As I am preparing to go on air, I'll imagine the camera is a person. I'll let my feet and body settle wherever they'd be if it were a person. I'll use my hands as if it were a person. It instantly shifts my delivery away from a more formal style and towards something more natural and engaging.

One last thing: movement. For years, I presented with a large touchscreen that is parked on the mezzanine looking down into the BBC newsroom. It's so big that to use some of its features I had to walk to them. Learning to do this taught me a valuable lesson: that movement is fine, but you need to know where you want to stop.

The problem is never the movement, the problem is if you don't know where you're going or where you're going back to. If you're wandering aimlessly it can convey a lack of confidence or purpose. If you look like you know exactly why you're moving and where you're going, then it doesn't do any harm at all and, in many cases, can add purpose. Wherever you are talking, if you have a space in which you can move, take time beforehand to consider where you're starting and where you might want to go. Practise if possible. You'll soon work out the places where you feel comfortable speaking from and where you don't. What we want to avoid is ending up in a no man's land – where you are left floating in the middle of a stage or room and you're not quite sure why or how you ended up there! You'll notice it and it may affect your delivery. The audience will notice it too and if they sense a lack of certainty it may impact their confidence in what you're saying.

Even if you are standing in front of a group and, largely, are going to stay in the same spot, some movement is fine. You don't need to do an impression of a statue. Allow yourself to move your feet and body just as you would if you were talking to someone in a conversation.

As with so many aspects of delivery, it is less about there being a correct thing to do and more about you deciding what you are going to do and being clear on the plan.

We are now on the home straight of our preparations to deliver our explanation. There remains one final part of the process. It's the part that ties everything together.

THE IMPORTANCE OF REHEARSAL

Beyond how much time you have, there's no real limit to the benefits of rehearsal. Do it enough and you will reach a point where your explanation feels clear and efficient, where the words belong to you and where saying them feels familiar and comfortable. That's a very good place to be. Rehearsal gets you there.

Here's a rehearsal checklist:

- Run through each section in turn.
- Run through the entire explanation.
- If there were parts you weren't happy with, go back over those.
- Are you comfortably within your time limit?
- Recreate the environment you're going to be speaking in as best you can. Including how you might need to walk to a position or sit behind a desk. (You may feel silly doing this, but don't let that stop you. It really helps.)
- Do you want to record it? (You won't always want to, but for explanations that really matter, it's worth

putting aside any discomfort you may feel about this and doing it. No one needs to hear the recording. You're certain to spot aspects you can improve.)

QUICK CHECK

- Is there any aspect of the explanation and how you're going to deliver it that you're still not sure of?
- Have you rehearsed what you're aiming to do?

YOU'VE COMPLETED THE SEVEN STEPS

This is a moment of great satisfaction. A lot of work has gone into getting to this point. In return, I'm hoping you feel very well set to give a great explanation.

It's always been important to me to connect the granular work of refining how we explain with the deeper purpose of understanding *why* we're doing it. In the Seven Steps, I'm asking you to consider a lot of different aspects of how you communicate. Some of them will become automatic; some you'll still need to go through systematically. For me, this work stopped feeling like an optional extra when I understood how fundamentally it could impact how I communicated. Or to flip that round, once you become aware of everything you do that gets in the way of what you want to communicate, it becomes hard to ignore it. If you see yourself saying and writing things that actively undermine how well you make yourself heard, I'm sure you're going to want to deal with that. I know I do.

The detail of how to do that is in the pages that you've just read, but if you need a quick reminder, here is the short version of the Seven Steps again:

STEP SEVEN

SEVEN-STEP EXPLANATION – QUICK REFERENCE

1. SET-UP
2. FIND THE INFORMATION
3. DISTIL THE INFORMATION
4. ORGANISE THE INFORMATION
5. LINK THE INFORMATION
6. TIGHTEN
7. DELIVERY

You're all set. Good luck!

4

SEVEN-STEP
DYNAMIC
EXPLANATION

The pursuit of clarity can sometimes feel like a particularly painful effort to get fit. Session after session you keep coming at it while wondering if it's ever going to feel better. Eventually, if you've stuck at it, you reach a point where what was once out of reach and uncomfortable feels within range and enjoyable. To be honest, that's a rare feeling in my sporting endeavours, but with explanation I know that feeling of breaking through. What felt overwhelming, feels tamed. What felt infinite, feels contained.

This is hard enough in controlled scenarios. When we write an explanation or prepare a spoken explanation, everything we're communicating is within our control. We decide exactly what we're going to say. But there are many scenarios where we're not in control. In fact, this is surely the majority of scenarios. Because every time we interact with someone else, we don't know what they'll say. We can't decide the questions we're asked, the reaction we receive or how well someone engages with us. These fluid scenarios offer us a further challenge.

How can our explanations meet all our expectations – precise, clear, efficient, high impact – in situations we can't control?

To do this, we need all the techniques we've looked at already – but we're going to layer another set of skills on top. Because the more we look at fluid and uncontrolled situations, the more we'll find that we can, in fact, control many aspects of them. I first realised the degree to which this could be done back in 2002.

I'm at risk of sounding melodramatic if I say a six-hour BBC training course was one of the most consequential days of my adult life. But leaving some of the events that would obviously trump that – meeting my wife and our children being born being the clear front-runners here – this really isn't hyperbole. Here's why.

This course took place in 2002 and, as I remember, was called 'Controlling your two-way'. A 'two-way' is broadcasting jargon for when a presenter and a reporter discuss a story. In its most basic form, the presenter says something has happened and then asks the reporter to tell us more. You'll have seen this hundreds of times. When watching, you may also have thought, 'Do they know what they're going to be asked?' The answer, much more often than not, is that they don't. The reporter will know the story they're going to be asked about – but beyond that they'll be ready to take what comes. What was always striking to me as a viewer was how that didn't seem to put reporters off their stride. I was about to start understanding why.

The course looked at two-ways from the point of view of the reporter. The trainer introduced himself and said, as best as I can recall, 'Today we're going to learn to say what you want to say regardless of what you're asked.'

'Really?' I thought. 'How does that work?' I'd been on the radio a few times by this point but had always considered myself entirely in the hands of the presenter and what they asked me. Given how intimidating I'd found the experience, the thought that I might have some control over what I was talking about was very attractive. I was seduced by the idea while being decidedly sceptical of it too.

The course was focused on a breaking-news scenario. Over the next few hours, we worked first on how to organise the basic information that we wanted to share. Then we looked at how to make sure that we got that information across, whatever questions we were asked. In this case, the training was aimed at helping us navigate a two- to three-minute radio appearance. This was an important skill for me to learn. My head, though, was somewhere else entirely. For me, this was like stepping through the wardrobe into Narnia. A whole world of possibilities opened up to me that went far beyond doing a two- or three-minute report, and beyond journalism itself.

I started to study much more carefully how conversations, interviews, discussions and meetings worked and how we all move the subjects of conversations around. I started to observe how talented communicators shared the best information they have within regular conversation. I started to notice how high-impact and well-crafted phrases were being used in fluid and informal situations. While some of this was undoubtedly happening by chance, that course in 2002 opened my eyes to the fact it didn't have to happen by chance. The more I looked, the more I saw ways that I could improve the clarity of my communication in conversations, in meetings, in correspondence, in interviews, in negotiations – to be honest, in an almost infinite number of scenarios.

Not only that, it gave me the confidence that my approach to distilling and explaining complex information in controlled situations – say, a university essay – could help me in fluid situations too.

In the years that followed that course, I started to really focus on how I organised and used information without notes or scripts or any prompts at all. At first, it was very much a work in progress. But the seed that the 2002 course had planted continued to grow. I noticed what was working and what wasn't and kept refining how I did it. That day had changed how I saw this aspect of communication.

Needless to say, I've not mastered it yet but I have had to come through any number of unpredictable situations where I know I'm going to need to talk about a vast subject without notes for a great period of time and found ways of that not being terrifying. Or, to be more positive than that, it can be exhilarating to know you are seeing a subject so clearly that, with various techniques at your disposal, you can take all comers and be confident of delivering a clear, detailed and engaging explanation.

The sensation is perhaps a little like flying – a huge amount of energy, noise and turbulence to successfully get off the ground and gain altitude, followed by something that is far calmer and controlled.

You can probably think of times when you're watching a TV presenter or livestreamer talk fluently at huge length about a range of subjects and you're thinking – how did they do that? Well, each of us has our own techniques but what I'm about to share with you is essential to me being able to do it. What at first glance appears impossible, becomes manageable – even, dare I say it, fun.

What follows here isn't about being a TV reporter (though if you aspire to be one, hopefully it is helpful). This is really about how we all explain ourselves in the most demanding scenarios. Here goes . . .

These are our Seven Steps – but this time with a twist.

1. SET-UP

2. FIND THE INFORMATION

3. DISTIL THE INFORMATION

4. ORGANISE THE INFORMATION

5. VERBALISE

6. MEMORISE

7. QUESTIONS

STEPS ONE TO THREE: PREPARE THE INFORMATION

The start of the process is exactly the same as before. I'm not going to go over that in detail again as you can easily refer back (see pages 100 to 121).

Just as in the earlier version of the Seven Steps, you'll now have a list of essential pieces of information. And just as before, at this stage it won't be organised in any particular way. We'll put that right now. But we've got to be even stricter than before.

STEP FOUR: ORGANISE THE INFORMATION

This is the first point in the process where we really start seeing a difference between how we'd approach a 'controlled' explanation and a 'dynamic' explanation. With the former, it's possible to keep a considerable amount of information within each strand, but with the latter we are going to have to slim each strand right down. The reason being a simple one: if you are constructing a presentation or an essay, you can piece together all those elements in your own time. If you're answering questions verbally, you are doing the construction there and then. Your building blocks need to be simple, memorable and usable. To that end, I am going to limit us to five pieces of information per strand.

Depending on the size of the subject you're working on, you may have three strands or thirty, and it may take you minutes to prepare or days – either way, for each distinct section of the subject, let's aim to keep the number of elements to five. More than five and it becomes very difficult to memorise and, just as importantly, very difficult to access in the moment. If you feel that you really need to use more than five, then think about splitting this strand into two.

This is the structure I always start with.

STRAND A	STRAND B	STRAND C	STRAND D
Primary point	Primary point	Primary point	Primary point
Fact	Fact	Fact	Fact
Fact	Fact	Fact	Fact
Fact	Fact	Fact	Fact
Context	Context	Context	Context

In each strand, I like to be completely clear on the point of clustering this information together. If I were to say this collection of information out loud, what would I be trying to achieve? What is the single purpose of doing that? Sometimes this will be very obvious, sometimes you'll need to ponder it. Either way, it is vital you have clarity on this before you go any further. That clarity will pay you back when you're trying to stitch this information together in real time.

Below your 'primary point' you'll have some facts that accompany it. We also need to provide context to go with it. As we've already considered, without context, information loses much of its meaning and impact. Where possible, add a line about this.

Here's a simple example if you were going for a job.

- **Strand:** Organising large events
- **Primary point:** extensive experience across event types and countries
- **Fact:** Project-managed a 2,000-person conference
- **Fact:** Responsibility for budget, personnel and marketing
- **Fact:** Freelance event organiser in five countries
- **Context:** Recently promoted to head of events at current company

You might look at this and think, 'I have more to say on this.' This is something for you to gauge. If, to continue with this example, you were going for a job that is all about events, each of those 'facts' could become strands in their own right. However, if the interview could touch on a range of necessary experience of which events is only one small aspect, then what you have here is enough. This is why Step Two is so important. Your assessment of the information you need will decide how you organise the information you have.

The other important point is that you don't need to write down everything you have to say. Take 'fact one' on the conference, once you've recalled that you need to talk about that, you will naturally, if you've time, be able to expand on it.

In some cases, four strands will be sufficient. In others – an interview, a staff Q&A, an industry gathering – you may well want to have more. If you have a lot, it can also be useful to group the strands so that you've thought about which ones connect closely together. This will allow you to move from one strand to a related one more easily. For example, if you're asked about 'managing others', you know straight away the two strands that will help you answer that. If we continue our interview example:

Managing people	Strategy	Innovation	Sales
Strand A	Strand B	Strand C	Strand E
Strand D	Strand G	Strand F	
	Strand H	Strand I	

STEP FOUR

Now we're in a good place. We've decided what we want to explain, found the information we need to do that, distilled it, organised it into strands (with a clear purpose to each) and, if need be, organised those strands into groups too. *A lot* of hard work goes into that. It's worth it. Many situations require us to access and communicate large amounts of detailed information. Doing this work in advance will vastly improve your chances of doing that really well.

Now, for me at least, the fun really begins.

QUICK CHECK

- Are your strands of information clearly defined?
- Are you happy with the information you have within each strand?

STEP FIVE: VERBALISE

In the 'controlled' version of the Seven Steps, verbalisation comes last. In this version, it comes now. Because instead of writing (or 'linking') around the information, instead we are going to try to use the information in an improvised and verbal way.

Before we get into the 'drills' needed to familiarise ourselves with the information we've prepared, it's important to explain why verbalisation is so crucial to this kind of explanation.

Earlier, I talked about the importance of familiarising our brains and our tongues with the words we're going to say. That's certainly the case with planned speeches and presentations. With dynamic explanation, it's not just important, it's essential. If the first time you try to construct verbal explanations is when you're doing it for real, the chances of it going well are really quite low in my experience.

My layman's reading of why this works is this. In any interaction with another person or group of people, my brain is having to do a few things: it's working out what's being asked of me, working out how I can respond, working out which parts of my knowledge to share and how to share them, gauging how much time I have to do that and how much detail the people I'm speaking to require on this subject. That's a lot for my brain to be busy with!

I can make all that a lot easier by front-loading some of the calculations in advance. If I do this, I can take as much time as I like beforehand to make good decisions about what to include and how to link the information together. Also, if these decisions have been made, I'll have enough brain space (for want of a more technical phrase) to evaluate the moment that I am in and the questions I'm being asked.

I think of my mind like a processor in a PC. If you run too many different programmes on a PC, it slows down and all the programmes underperform. If you close some programmes, the performance of the others improves. In a similar vein, the more I've got my mind used to the information I want to use, the more capacity I have for decisions in the moment. That's the theory anyway!

Crucial to doing this well is verbalisation. For me it's the differentiator. Done well and methodically, it elevates my ability to explain beyond anything I could do by preparing written notes and learning them by rote. It's a lesson I've learned a few times over, not least when I was dispatched to cover the Dutch elections of 2017.

Next to the Dutch parliament in The Hague is a square called Het Plein. On one side is the parliament complex itself, on another cafés have their tables set out under umbrellas, and, on market day, stalls selling books, food and clothes surround a statue of William I, Prince of Orange.

In 2017, this square was my home for a few days as we covered the election campaign and election day itself. On the first morning of broadcasting, I was waiting for my first 'live' to come around. I'd done any amount of preparation and felt good to go. Except I wasn't. Talking to a colleague about the campaign I was nowhere near as fluent or clear as I needed to be. With the help of our producers, I'd done a perfectly good job of organising the information I needed. On my computer, the preparation looked as good as it could be. What I hadn't done was verbalise the information enough. Nowhere near enough.

For me, it's as if my mouth and brain need to get used to using the words. Somehow it helps to create a network of connections around the information and to situate words and phrases around that network. Verbalising makes the information you have much, much more usable.

That morning, all was not lost. I still had some time to spare and so, with my notes on my phone for reference, I walked around the square talking to myself. Goodness knows what the traders setting up that day's market thought as I wandered by reciting coalition government combinations, but I didn't let feeling a little daft get in the way of doing it. I'd think of a strand of the story and try to talk through it. It was worrying how bad I was at first but, within minutes, where I'd been stumbling, I flowed, where I struggled to explain the relationship between different elements, now I could do so coherently.

I see other reporters and correspondents doing this all the time too. I'll sometimes be doing a stint in Downing Street on the latest political drama. On those days, the street can be quite cosy as TV journalists gather close in order to have the famous black door over their shoulder. It's not unusual to see one journalist or another quietly talking to themselves.

In our own ways, we've all worked out that if the first time you say something is the time that it matters, you're taking a chance. Only when you speak will you really understand which phrases work, which points work, which ideas you're comfortable handling and which names you're confident using. I know I need to try out what I've learned if I'm to make the most of it.

To help me do this, I developed a series of 'drills'. I find them essential if I'm without notes. But before we launch into those drills, a word about notes themselves.

During my time as a presenter, I'll have done hundreds if not thousands of 'two-ways' – those moments in the news where the presenter talks to a reporter. If a reporter is joining me on the set of my programme, most will arrive with either no notes or a single page of them. Sometimes, though, a less experienced colleague will arrive laden with pages of notes. They're anxious they may forget an important detail. I know that feeling too. And there have been times I've shown up with lots of notes.

Even now that I'm more experienced, if I'm reporting on a complex story, I may still bring a single page of well-organised notes. And here's what almost always happens – I never look at them.

Notes can be useful if there's a specific quote, statistic or fact that you want to be sure of. In a breaking-news environment, reporters will refer to notes a lot and that works both for them and the audience. But in most scenarios, such as a class discussion, a conference panel, an interview or a client meeting, notes are very unlikely to help you. It's just not possible to keep looking down, to find different pieces of information according to what you're being asked *and* to talk engagingly at the same time. At least, it's beyond me, and I've seen a lot of other people coming unstuck trying.

If I'm joined on TV by a reporter or guest who may be nervous, I suggest that they come a few minutes early. We stand on set and just chat about the story. This gets them used to the environment (which helps, as we've already discussed) and gets them flowing on the subject. After we've chatted, I always suggest not to worry about the notes, or at least to pick the most important page and just leave that on the table. And I say, 'Please keep on talking just as you have been doing.' I'm not doing this to be unsupportive or to deliberately give them a journalistic baptism of fire. Far from it. I'm doing it because I know *not* having pages of notes is going to make their life easier, not harder. If the notes aren't there, they'll talk – almost certainly superbly. If the notes are there, they'll look down, invariably not be able to find the words they're looking for and lose their flow. As we're about to see, notes are very useful as you prepare to explain yourself verbally. But when it comes to actually speaking, they can create more problems than they solve.

What can really help are the following drills . . .

VERBALISE EACH STRAND

Step by step you're going to get used to using the information that you've prepared. And each step is going to involve saying this information out loud.

Try talking through one of the strands

Imagine someone has asked you an open-ended question about the subject of this strand. Allowing yourself to look at the information, try talking through it. Does the order of the information work? If not, switch it around. Try doing it again. If you're happy with how you joined two of the elements, make a mental note of the phrases or language you used and resolve to keep using them. If one join lacks precision and fluency, experiment with how you could improve it. Try the whole thing again.

As you do this, the whole strand may take less than a minute, or it may take a little longer. We want how you talk on each strand to be precise and high protein. We're less concerned about its duration.

Do this with each of your strands. Some you'll get very quickly, while others will take time. You might find you're missing some information, in which case go back to your original list of distilled elements to see if you can fill the gap. And all the time, check that what you're saying supports the 'primary point' you've chosen for each strand.

On many occasions, you'll only have three to five strands and this won't take too long. This can be excellent preparation ahead of regular meetings or conversations where you don't need to share a lot of information but still want to make sure you say it well. On the occasions that there are more, often these are subjects you may talk about repeatedly. If it's for an interview, it won't be the last time you need to explain your achievements, skills and ideas. If it's about a major piece of work you're involved in, you may use these explanations again and again. I did a lot of work in the early days of the 50:50 Project refining my explanations for dynamic scenarios. Since then my colleagues and I have used those explanations hundreds of times. The kind of clarity and fluency you gain from working through this process will pay you back, I promise.

Connect two strands together

Next, pick any two strands and talk through one and into the other. Think about how you shift from one to the other. Once more, if you do it well, make a mental note. If you don't, how can you do it better? We're aiming for you to sound focused and informed but also conversational and fluent.

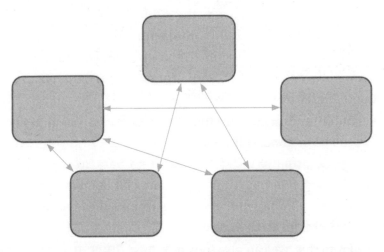

Practise moving from strand to strand in different combinations.

Keeping picking out different strands and talk through one into the other. Sometimes they may complement each other, other times they may be disconnected. For the latter, explore how you can shift from one to the other while maintaining your fluency on each strand.

The way we shift from one section of pre-prepared information to another is absolutely crucial to delivering dynamic explanation. You can't really practise this too much. And central to your success will be having a range of bridging phrases at your disposal.

Bridging phrases

Earlier, we looked at 'joining phrases' that move you from one element to another within a controlled explanation. These can be specific to the elements because you are choosing the order in which everything comes.

With dynamic explanation, we don't know the order in which we'll be using our strands. As such, getting from A to B or from C to A smoothly is vital to constructing explanations in real time. 'Bridging phrases' perform the same function as joining phrases but are generic. They don't contain any specific content, but they allow us to manoeuvre between strands with minimum fuss. Here are some of my favourites.

- That's one area I'd emphasise, another is . . .
- There is, though, more than one aspect of this issue to consider. Another is . . .
- Another thing I'd stress is . . .
- And while that's important, so is . . .
- This also links to . . .
- From this, we can also look at . . .
- There's more than one dimension to this. Another is . . .
- But to understand this issue, we can't only look at this – we also need to look at . . .
- And that connects to . . .

- And if X is one aspect of that, Y is another . . .
- There are, though, several ways of explaining this. Another is . . .

As you can probably tell, I'm a bridging-phrase enthusiast. These are just some but there are many more. You'll come up with your own ones too. They all, in their different ways, mark that an aspect of your explanation is finished and move you on to the next.

Try using some of these phrases as you talk through one strand and into another. Looking at your notes and a list of bridging phrases is fine. This part of the process is all about fluency. The more you do it, the more fluent you will become. As well as that, the more you do it, the easier memorising all this will be when the time comes.

There is, though, one more thing that, right in the moment, can help you select the best information to use.

IN SEARCH OF TIME

If we're to successfully explain ourselves in fluid, unpredictable environments, we need time to think.

In a situation where you know which strands you're going to talk about, you're already going to be excellent either with notes or off by heart. Here, though, we don't know what we're going to be asked and would appear to have no time at all to line up what we want to say.

Finding time to get your thoughts organised in this situation may not feel at all realistic. But within those interactions are pockets of space that we can make the most of.

In part, this is about knowing where that space lies and having the confidence to use it. In part, it is about creating space for ourselves. And if, in the middle of the most important and pressing of interactions, you can successfully create space for yourself to think, the effect can be incredibly powerful. It can feel like playing the game by different rules. With more time, we can make better decisions about how to respond to what other people are saying. We can convincingly match what we're being asked with what we want to say. It's hard to overstate the impact this can have on the confidence, calmness and clarity with which you can talk.

My discovery of 'time' as the key to dynamic explanations was initially inspired by one short passage of a book that I read on a holiday.

Matthew Syed's *Bounce* came out in 2010 and I think I read it that summer or the one after.[10] It's a book about sporting success and performance and the degree to which innate talent is a factor. (Syed's thesis is that 'talent' is much less of a factor than is sometimes claimed and that, in fact, purposeful practice is more significant.)

A few pages in, there is a section that had such an impact on me that I've remembered it ever since.

The section begins with Matthew Syed describing a promotional day with various people invited to play a little tennis with former Wimbledon champion Michael Stich. I'll let Matthew Syed take up the story. Bear in mind as you read that Syed was a table-tennis professional before becoming a successful writer.

> I asked Stich to serve at maximum pace. He has one of the fastest serves in the history of the sport – his personal best is 134 mph – and I was curious as to whether my reactions, forged over twenty years of international table tennis, would enable me to return it.

Michael Stich agrees. He leaves to warm up and then returns.

Stich tossed the ball high into the air, arched his back, and then, in what seemed like a whirl of hyperactivity, launched into his service action. Even as I witnessed the ball connecting with this racquet, it whirred past my right ear with a speed that produced what seemed like a clap of wind. I had barely rotated my neck by the time it thudded against the soft green curtains behind me.

Matthew Syed's table tennis experience had made no difference.

To understand this, he visits an academic at Liverpool John Moores University called Professor Mark Williams. First, we learn that professional tennis players look at the body to read the serve (I'd been looking at the racquet which, I imagine, is just the start of where I've been going wrong . . .). Professor Williams explains:

> It is not as simple as just knowing about where to look; it is also about grasping the meaning of what you are looking at. It is about looking at the subtle patterns of movement and postural clues and extracting information. Top tennis players make a small number of visual fixations and 'chunk' the key information.

In other words, the professionals know what information to look for and know what to do with that information. That gives them a significant edge.

Then came the bit which, over time, has come to change my thinking about explanation. Matthew Syed quotes Janet Starkes, professor emeritus of kinesiology at McMaster University in Canada, who says this:

> The exploitation of advance information results in the time paradox where skilled performers seem to have all the time in the world. Recognition of familiar scenarios

and chunking of perceptual information into meaningful wholes and patterns speeds up processes.

I loved all of this. I've always played and watched a lot of sport and often wondered why some players appear to have more time. I liked too the idea of the importance of recognising 'familiar scenarios' and how that can speed our responses to them. It all had a great impact on me – though, at the time, I confess almost entirely in terms of how I thought about excellence in sport (or a lack of excellence in my case . . .).

That was 2010 – just around the time that I started presenting regularly on TV. In the years that have followed, I've faced many different situations where I've needed to be able to explain myself clearly and persuasively, without notes and often with the verbal equivalent of a Michael Stich serve coming my way.

It was fascinating to go back to this passage as I write this book. While I remember and reference the Stich story regularly, I am pretty sure I've not reread it since 2010. Going back through it, the relevance to explanation is there at every turn – from its wider ambitions to understand how we create time and space for our brains to make decisions, to how we can organise and learn information in advance of being in dynamic scenarios, to even the use of the word 'chunking.'

When I read that passage in 2010, I didn't understand how the concepts of memory, 'chunked' knowledge and time connect so powerfully to communication and explanation. However, in the years that have followed, I started to make that connection. Just as the best tennis players can both read a serve to give themselves time and then, because of their training, have the right response to it, so, in conversation, we can spot what is being asked of us, create space to think about how to respond and be ready with the right information for that moment.

My pursuit of time led me to the final two steps in my approach to dynamic explanation.

In Step Six, we'll look at how we can memorise information and how best to use it.

Step Seven is focused on predicting which questions we may be asked and, in the moment, assessing what information is required of us and how best to explain it.

All being well, both will give us the space we need to think as clearly as possible.

> **QUICK CHECK**
> * Have you verbalised your information so that it feels comfortable to say?
> * Are you comfortable moving between different strands of information?

STEP SIX: MEMORISE

A few years back, I was sitting in the living room of the eldest of my two sisters. We both had young kids and there were toys and other child-related paraphernalia all over the place. We sat amongst all of this having a cup of tea and trying to catch up. One toy caught my eye. It had four buttons: red, blue, green and yellow. I recognised it straight away. It was the memory game 'Simon'. If you've not had the pleasure, when you start, one of the buttons flashes, you press it. Then the same button flashes, followed by another. You then have to press both in the correct order. The machine then flashes three times, again you repeat the order. On it goes. That's it. That's the game. The challenge is to see how many consecutive flashes you can remember.

During that visit to my sister, I became very taken with this. At first, I found it hard. Maybe getting to ten flashes and then stalling. Then I noticed that as my brain grappled with the task, it was beginning to automatically group the flashes. I'd remember 'green yellow yellow green' as a block and then 'red blue yellow yellow green' as a block. I couldn't always quite work out the reasons why my brain would favour ending a block at a certain point, but I didn't worry about that too much. The blocks were making it far easier to remember the entire sequence. So long as I remembered that a block existed, I never forgot the colours within it.

At this point, I'm not sure I was contributing quite as much to the conversation as I might have been. I put the toy down and turned my attention back to my sister and the kids. But it had got me thinking.

After I left, I looked up 'Simon' in the app store of my phone and, sure enough, there it was. One download later and my efforts to set a new record continued. The better I got, the more I noticed something else was happening. I was already thinking of the blocks as single entities rather than as a run of five colours (or however many were in it). Now I was starting to remember blocks of blocks. I was also creating associations with each individual block ('double green', 'yellow sandwich' or whatever silly name came to mind) and then I was remembering them all together as a single entity – 'double green', 'yellow sandwich' and 'run of blue' would become the 'First Block'. So long as I remembered there was a First Block, the order of individual blocks within that came quite easily and the colours within each one always came. I was intrigued to know what was going on here. I was able to bank a lot of the hardest memory tasks so I could concentrate on remembering the newest colours as the machine served them up.

Not long after I rediscovered 'Simon', I was asked to go and cover the Greek debt crisis in 2015. I talked about this earlier in the section on 'Complexity'. The trip posed quite a challenge. This was a sprawling, complicated and rapidly shifting story. It was also the lead story for much of the ten days or so that I was there. That meant going on air a lot. As I mentioned earlier, there's too much going on to find out what you're going to be asked in all the different live reports. It was also the first time I'd done this kind of reporting and presenting on a big story for TV. I'd done it many times for radio, but radio has one big advantage – you don't need to look at the listener as you explain the story. If you feel the need, you can have a computer or a piece of paper in front of you with all the information and themes you would like to share. On TV there's no such option. It's fine to look down occasionally if you want to check a fact or a quote, but most of the time you need to be looking 'down the barrel', as we say. In other words, the vast majority of what you say will be off the top of your head. As I wrestled with the latest policy shifts of the European Central Bank this was not without its challenges.

Doing my best to escape the unremitting Athens sun, I started experimenting with ways of not just taming all the information I needed to reference but also being able to recall it at very short notice. I thought about how my brain was helping me play 'Simon'. Despite Professor Williams using the phrase 'chunk' in Matthew Syed's book, it wasn't something I was familiar with. But what I was doing when playing 'Simon' was, it turned out, chunking. I was clustering information into one chunk and then remembering that chunk as one entity rather than all of the elements within it. I started to try similar tactics with what I wanted to say about the situation in Greece. To my excitement, it worked. It made navigating those ten days or so in Athens much easier.

On getting home, I immediately wanted to know more about how memory techniques could help me. I especially wanted to explore how better to access different chunks at different times and how to take decisions about ordering them. Because there is a difference between playing 'Simon' and what I needed to do in Athens. With 'Simon' the order stays the same; with dynamic explanation it doesn't. My situation wasn't only about linear memory; it was also about what I call *access* and *rapid ordering*. To explain yourself with clarity in dynamic scenarios, you need to be able to access chunks of pre-prepared information very easily and then order them according to what you've been asked.

As I've realised in the years since, this is something that is helpful well beyond standing in front of a TV camera in the midday sun. Being able to access and organise interconnected information by memory is invaluable. I use this in a lot of day-to-day exchanges, such as work meetings or a conversation with my kids' teachers or briefing a tradesperson, as well as for public speaking, Q&As, more formal meetings or making a pitch. You'll be able to think of situations in your life where you need to have information organised and ready to go – but can't have it written down in front of you.

If you can do this, it can give you a significant edge. It allows you to be fluent and precise across a range of detailed information – despite being in a dynamic environment.

For me, when memorising goes well, it's almost as if I can see the information as shapes. It becomes like an array of kids' building blocks that can be picked up and arranged in whatever order is required. I still need to decide what I want to make with them but the blocks themselves are there and they're ready to use.

In the years since Athens, I've worked hard to hone how I do this. This is where I've got to.

MEMORY TECHNIQUES

There is no one-size-fits-all approach to using memory to help you communicate. Just as with explanation more broadly, which technique you choose will need to match the situation that you're facing. But for me, chunking is at the heart of all the ways that I use memory techniques.

At its simplest, chunking involves identifying a collection of information, learning its order and then labelling it as a single thing in your mind. The next level of chunking is to order the chunks themselves and then remember that as a single entity. This is essentially what I was doing without realising it when I was playing 'Simon'.

To make the most of chunking, we need to go back to our drills from Step Five. The information we have for each strand is a single chunk; I'm effectively using these words interchange-ably when talking about memorising.

The last time we went through the drills, we did so with notes. Soon it'll be time to put the notes away.

Level One: Single strands from memory

This level and the next are for the quickest of meetings or con-versations where you just need to make sure you get certain things across.

We need to start testing how well the strands roll off your tongue without the help of your notes. If you did plenty of practice during Step Five, you may be in for a pleasant surprise. Sometimes turning the notes over can be the equivalent of a child moving from a balance bike to a pedal bike – they often ride first time. You may find you can deliver each strand just as easily without the notes without any extra effort.

The overall number of facts, points, pieces of context and questions that you want to include will vary. Either way, you don't want more than four or five elements per chunk. You'll struggle to remember them if you go for more.

Let's imagine your strands look like this.

STRAND A	STRAND B	STRAND C
Fact	Fact	Fact
Fact	Fact	Fact
Point	Point	Point
Context	Context	Question

A good starting point is not to talk them through as you would to someone. Instead, just recite the elements like a list. Take each in turn and see if you can punch out the information that you've learned. If you can, do the same but this time linking them together in normal speech.

Level Two: Two strands from memory

This is a straight repeat of the drill you did with your notes in Step Five. This time, pick two strands that complement each other, check your notes if you need to, then put them to one side and talk through one strand and into the other. If it goes well, see if you can do it in the reverse order. This second part is really important. We're not just learning to repeat each individual strand – we're also learning how they fit together.

Keep selecting pairs. Be quite exacting in your standards. If you feel yourself waffling as you work through the information, go at it again. You'll notice that the extraneous words you were adding start to disappear. The efficiency with which you move through the information will improve. What at first felt awkward can feel comfortable and clear.

When you feel ready, write the names of the strands on to pieces of paper and select two at random from the pile. Just as before, see if you can run through one and transition to the other.

Level Three: Multiple strands from memory

This is for lengthier meetings, interviews and conversations of all kinds.

If you have fewer than five strands (or chunks if you prefer to call them that), decide on an order and try talking it through from start to finish.

Don't fixate on the order. You may have no choice but to discuss this another way around.

Next, talk through the five strands again, each time picking a different starting point. See how well you can work your way back on to your preferred order from wherever you start.

All being well, you'll start connecting the strands together with real confidence. We want everything you're doing – talking through the strands, moving between them, switching the order of them – to feel familiar.

This approach can also be used if you have far more than five chunks.

This is what I use in much of my reporting. It works in any dynamic scenario where you're going to need to access a decent amount of complex and varied information. This could be a job or university interview, a board meeting, a vital staff meeting, a Q&A at a conference or a meeting with a major client.

In these situations, the amount of information you're handling is sizeable so there are a couple of extra considerations.

First, you'll have more chunks – maybe between ten and twenty. Stay disciplined on not overloading each chunk. You won't remember all the elements if you do. Better to split a strand and create two separate chunks that have their own labels.

Second, think about how all these chunks cluster together in answer to different questions. On one subject, you may want to use two or three, on another perhaps three or four others. If you want to practise, write out some of the questions that you're expecting. Pick them out at random and see how the ordering of the chunks that you've planned feels as you say them out loud.

You can't be rigid about this ordering because every question may require a different start, middle and end to your answer. But having familiar ordering patterns is very useful. When you see a certain question, you'll think – 'Ah great, I'll do this followed by this followed by this.' In those situations, by the time you go to speak, your work is already done. As I'm writing this, I can think of European Union summits in Brussels where I'd done so many of these drills that as a question was coming my way, it was as if I was placing the chunks on a conveyor belt. When it came for me to answer, I could see both what I was about to say and what was coming next, and chunk by chunk my answer would play out naturally.

This part of the process is hard, though. As you experiment with different practice questions, you'll be working out new orders too. If it doesn't come out as you hope, that's completely normal – it rarely does first time. But bit by bit you'll settle on orders that work and that you can deliver.

There's no limit on how much time to put into this. The more you do it, the more easily you'll remember what to say and the more precise you'll be with the delivery of it. That's a great feeling. But if you've time, don't stop as soon as it starts to feel manageable. If you've time, push on from there until how you talk through these chunks and how you organise them becomes close to automatic. That is when you are really ready to use memory to transform how you communicate.

I should add that if you're doing this for a major piece of work, this may take several days or even weeks to get right. It definitely doesn't need to be done in one go.

All this puts you in a really strong position. Just one of these chunks is going to make for a detailed and thorough answer. Put a series of them together and how you're sharing information will be structured and high impact – as well as being rich in detail.

Level Three is as far as I'm going to take it. But there are even more sophisticated ways to help you remember what you want to say – and there are people who are far better at it than me. Next, I need to bring in the expert . . .

Level Four: Lengthy and complex explanations

The moments when you need to do this may be fewer than some of the examples above, but these moments may also be hugely consequential. Perhaps you're revising for A-levels or university finals, perhaps you want to give a keynote speech without notes, perhaps you simply need to be across a large and new subject for a vast array of situations.

Dominic O'Brien is a multiple World Memory Champion who's written several books about the impact that memory techniques can have. He argues that improving how we memorise information 'reveals the very essence of the learning process'. It does so, he says, by 'strengthening your working memory capacity'.

I asked Dominic to explain more about what that phrase means.

'You can only hold about four to six bits of information in your short-term memory. If you go to a party, after you've been given four or five names, you're going to struggle. But if you use your working memory, then you're breaking names down into syllables and finding associations.

'Your working memory uses the information you've just been given and then processes it with information from your long-term memory such as places you've been to.'

This is the key point. We all have memories that are entirely set and which we are very unlikely to forget. If I asked you to describe your home, you would describe the layout correctly every single time. That is long-term memory. Memory experts advocate that we use that stable, zero-risk, long-term memory to support our ability to recall information.

Dominic quoted another memory expert, Dr Tracy Alloway, who describes our working memory as a desk. Some of us only have a small area on the desk where we can place memories, but it's possible to make that area bigger. This matches my experience of experimenting with memory techniques. I could feel my capacity growing as I learned where to store the information.

Dominic O'Brien now works with actors, CEOs, CFOs, comedians and others in helping them improve their ability to store and recall information, particularly when under pressure.

He says he teaches three main methods.

The first is the 'link method' (which you may also see called the 'story method'). This is where you have, say, five subjects that you need to touch on and you create a story – however fantastical – that works through those subjects in order. You retell the story to yourself a couple of times and the story becomes a single entity to recall. As you retell it, the subjects come to your mind in the order you need them. This, Dominic says, is best for smaller memory tasks and it resonated with what I've been doing on the TV. I am always looking for the 'link' that helps me remember to shift from one subject area to the next.

Now, in the case of the news, the link may naturally be part of the story that I am trying to tell. But sometimes there is no obvious link. Or you may have several unconnected subjects.

Dominic's suggestion here is to use a story to help with that. Having now tried this too, I can confirm it's incredibly effective. Not only does the memorising become a lot easier, it removes a lot of the pressure. The chances of you forgetting the story are really low. Suddenly what was previously difficult and unreliable, feels easier and certain.

That's the story method. But if you have a more complex amount of information to retain, Dominic points you towards the way he's won all his titles: the 'journey method'.

This is also a very clever idea. Here, you take routes that you already know well and attach whatever you want to remember to the route. Also, at the different locations on the route you can attach information to aspects of those locations.

For example, let's imagine you know your local park really well. The route you plot through it begins at the entrance gate, then past the bench you like to sit on, then to the playground where you take your kids, then to the tennis court you use and the car park where the ice-cream van is. Those five locations can be the five subject strands that you want to talk about.

Then, say, at the gate – there is a sign, a place to lock up bikes, a big tree and a bin. Each of those could be associated with pieces of information within the strand that you want to talk about.

Before you know it, you are remembering a raft of information and doing so entirely based on memories that you already have, and which are ultra-reliable. And lest any of us doubt how well this can work, Dominic routinely remembers thousands of pieces of information using the journey method.

But I had a question for Dominic. These methods are clearly fantastic for memorising planned information – a speech, a play, a presentation, an essay. But what if I don't know the order in which I need to access the information I've prepared?

'Ah,' said Dominic. 'This is where you use memory palaces.'

This fantastic phrase describes the most advanced of memory techniques. If you're a fan of the latest TV version of Sherlock Holmes you may already be familiar with the idea – though, in fact, the concept of 'mind palaces' goes back centuries.

This is how they work.

As with the journey method, you start with a familiar location. But rather than being focused on a journey along a familiar route, this time we're thinking of a place. It can be anywhere but I'm going to take the example of my house.

Each room is associated with one of the strands of information that I've prepared. If I was talking about world trade, maybe my kitchen would be America, my living room would be China, the study would be the European Union and so on. Then, just as with the journey method, within each room I'd associate familiar aspects of the decor and furniture with pieces of information. These associations may seem silly – say a record player in my living room might connect with a record level of Chinese exports. Dominic's advice is to take any association or connection that works for you, regardless of how absurd they may be. They don't have to make sense to anyone else.

Once you've done this, you can 'walk' around the memory palace, choosing which part of it to go into and allowing your long-term memories of each location to trigger the information you've attached to them. Being able to do this directly connects with our ambition to pre-prepare large amounts of information and then access it in the moment.

Choosing memory techniques

When I started messing around with memory techniques after my chance encounter with 'Simon', I saw them as a fun extra that I could experiment with. Now they're a core part of how I approach explanation. If you have gone to all the trouble of organising and distilling what you want to say, it's a terrible shame if you can't recall how best to say it.

Dominic O'Brien told me: 'I thought I was a dunce. If I can become an eight times memory champion so can anybody.' His point is that memory techniques can help everyone. They've certainly helped me.

If they go to plan as I prepare an explanation, I think of myself having a shelf with blocks of information sitting on it. Each block is one of the strands that I've pre-prepared. As I'm being asked a question, I look up to the shelf and think, 'Which of these shall I use?' According to what I'm being asked, I take one or two or three of them down from the shelf, put them in the order that best suits the moment and talk through them.

To do this well, though, I need as much time as possible to consider what I should take off the shelf. That I've already organised and memorised the information helps with this. We can also find time in the questions that we're asked, as we're about to see in Step Seven.

QUICK CHECK

- Are you clear on what you'd like to memorise?
- Have you decided which memory technique you're going to use?
- Have you practised your explanation without notes?

STEP SEVEN: QUESTIONS

In dynamic situations, nothing will do more to shape the opportunities you have to explain yourself than questions. They're an integral part of how we all communicate with each other. And if we're in a situation where we're the focus – say, a company briefing, an appointment with a client, a job interview, a conference panel and so on – then the questions will be constant.

In these moments, it's not you who's deciding when you speak or what the subject is. It can feel as if you're not in control. To some degree that is true – in the same way that Roger Federer is not in control of the serve that he receives. However, there are many ways that he controls *how* he returns the serve.

It is the same with explanation. Done effectively, the questions don't become irrelevant (far from it), but they cease to be the dominant influence on what you say. You can explain the information you believe is most relevant while matching it to the questions people have for you. It's good for you and it's good for them (this is where the tennis analogy ends, as clearly Federer's returns are often not good news for his opponents . . .).

PREDICTING THE QUESTIONS

One of the most powerful things I've learned from conducting hundreds of interviews and from giving hundreds of presentations and Q&As, is that, on the whole, people and issues are *predictable*.

I don't mean that to sound rude. We are all predictable. Our interests, concerns, turns of phrase, personalities and preoccupations tend not to radically shift from one day to the next.

If a person has engaged on a subject before, when they next engage on the subject, they may well say similar things. I'm living proof of this. If, for your sins, you had to listen to me talk about my explainer videos at five different events, you're going to hear me using very similar explanations, turns of phrase, statistics and ideas.

Not just that. If, say, one group of staff has some questions about an idea you're proposing, it's very likely another group of staff in a similar line of work will have similar questions. Or if you watch a politician, film star or businessperson do a 'round' of interviews one morning, they won't get asked identical questions each time, but some are certain to crop up several times in different guises.

If you're the one on the receiving end of the questions, this is all to your advantage. It gives you a good chance of guessing what is coming your way.

1. What are the questions you're very likely to get asked?

Think of everything you know about the scenario you'll be in – the likely subject, the people likely to ask questions, the likely areas of interest – and write down questions you think you'll get.

If the scenario is straightforward and short, this may be enough. But if this is a more robust or detailed arena, press on.

2. What would *you* ask?

This is one of my favourites. Imagine you were on the other side asking the questions. What would *you* ask? If you were being as curious and exploratory as you could be and were faced with yourself, what information would you like to hear? Add those questions to the list.

3. If you wanted to be awkward, what would you ask?

If you wanted to make life as difficult as possible, what would you ask yourself then? What are the most awkward questions you can think of? Add them to the list.

4. What questions would you rather *not* be asked?

This isn't quite the same as question 3. Sometimes, I'll be aware I'm avoiding talking about a subject because I haven't quite worked out how best to explain myself on it. Perhaps I've not quite found the right turn of phrase or I don't have the right piece of information to talk about it well. Note your discomfort! And add these questions to the list.

5. What questions at the periphery of the subject might come up?

There's always a chance you'll get asked something that is outside what is reasonably considered part of the subject that you're going to be speaking about. In my line of work on the news, this would be a small detail of a big story that I wouldn't expect to know. These are questions you really don't want to get but you just might. If you can think of some, add them to the list.

6. What can you find out about the people asking you the questions?

If you want to go the extra mile, this can be incredibly useful. There are many ways that we can find out about the people we're going to encounter. If you're meeting with or being inter-viewed by colleagues, you can ask other colleagues about them. What are their interests and passions? Which issues do they think are a priority?

If you're visiting a company or institution, you may be able to speak to someone there to find out its perspective on whatever you're talking about. Or there may be clips online to show you previous discussions featuring the people you're going to meet. See if you can spot a pattern in the issues or questions that are raised.

Maybe you're being referred to a consultant by a junior doctor. When you are referred, you could ask the junior doctor what the consultant is like and the kind of approach he or she takes to this illness? Or, as I've been writing this book, our eldest daughter has been doing exams. She looked up all the past papers for the exam board her school is using. That gave her clues as to the kind of questions the exam board favoured.

This process is not going to be perfect. You won't end up with a definitive list of questions that are certain to come up. But you can start making educated guesses.

Depending on the situation, by now you'll have a list that's anywhere between a couple of questions and a lengthy selection.

The next task is to split the list into three:

- Questions you think you can answer

- Questions you need to work on how to answer

- Questions you need new information to answer

PLOTTING YOUR ANSWERS

Let's begin with the first section of your list – the questions you think you can answer. If you have multiple questions that are essentially asking about one subject, settle on one and get rid of the rest. Once you've done that, it's time to get ready to use your strands. You've already prepared them and learned them. Now we're going to put them to work.

Just before you start experimenting with your answers, it's essential to emphasise one important aspect of dynamic explanation. For the moment, length is neither here nor there. Some of the questions on your list may be answered with 'Yes'. That may be all you need to say. Others may require a multi-faceted answer pulling in several of your pre-prepared strands. What matters is that what you're saying is relevant, efficient and clear. The duration is only important if your response is so long you think you may not have time for it.

Faced with the first question on your list, which combination of strands would you use to answer the question? Give yourself as long as you need to decide and then have a go at answering it. If it feels good, fantastic. If you feel happy with the content but it's not flowing, do it again and again until it feels comfortable. If you're not happy with the content, what other parts of the information you've prepared could you use? Bring it in and try again. We've got two goals here – for the information to match the question and for you to feel confident delivering it. When you get to that point, move on to the next one. If it's helpful, make a note of the strands that you used for that question.

Next on your list are the questions that you're not so sure how to answer. I'm hoping this is a relatively short list, as your work on Steps One to Four should have already tackled areas you weren't sure of. Take these questions in turn. What is it that poses the problem? Are you not clear on what you want to say? Or is it that you're not clear how to say it? Try to work out the former before you then move on to the latter.

Think about which strands of information can help you. Are you short of a phrase that could bring it together? If so, play around with what that could be. Doing so out loud can make you feel a bit of a lemon but keep going. The right phrase may well emerge. If you're still struggling, ask a colleague or friend or relative. Sometimes a second pair of eyes can help unlock it. Or if that's not possible, take a break. Your brain will keep thinking on it in the background. When you come back to it the next day, it may come more easily.

What is really important is that you don't ignore the fact that you're not comfortable on these questions. Much earlier we talked about how we can't avoid detail that is hard to explain. It's the same with hard questions. We really don't want to ignore them. For several reasons. First – the obvious one. If you get asked them, you want to be able to answer well. Given you're already struggling with them, the chances of this going well are low. Second, if you're aware that there are questions which you'll struggle to answer, that may impact your confidence throughout. Third, more likely than not this isn't the only time you'll get asked them. Getting them right will deepen your understanding of the subject and your ability to explain it. There are long-term benefits too.

Just as with the first part of your list, practise each response until you are happy with the content, precision and fluency. In particular, make sure you're happy with the phrasing you're using to navigate the answer. Often with difficult questions, even when we have all the information prepared, it is a turn of phrase that can both give the explanation its coherence and give you confidence to talk about it.

The last section of your list is for questions where you need more information. Now, we've already been through a process of identifying the most important information. I am going to assume we've not really missed anything obvious. These questions are ones that are some way off subject but that could come up. My rule of thumb here is that a little goes a long way. We can't prepare detailed answers to questions that are unlikely to come up – the work would never end! But we can try to anticipate where an unreasonable question may come from and have at least something to say.

Let's imagine I'm covering a big political story and there's been, say, a low-level resignation that's not of any great consequence. I might quickly check the name, the basic circumstances and what, if any, reaction there's been. I wouldn't be able to hold forth for minutes, but, if it comes up, I can fashion a quick answer before getting on to another topic as soon as possible.

On these questions, do a quick test that you have something to say on each – even if it's only a little.

By this stage, you are in seriously good shape to offer superb explanations when faced with all sorts of incoming questions and challenges. You've taken all the information you've so carefully organised and memorised – and are now well versed at using it when faced with a range of the most likely questions that you'll be asked. The familiarity of both the explanations and the likely incoming questions means that both will feel much more comfortable when you're doing this for real.

However, to make the most of your work so far, we need to add flexibility and spontaneity to our explanations. To do that we still need all the time we can get.

ANSWERING QUESTIONS IN THE MOMENT

We've already helped to give ourselves more time by doing all this preparation in advance.

By organising and memorising the information, that is all work that our brain doesn't have to do in the moment.

By practising the questions, we've already worked hard on how to answer them.

Just as a student who has revised an essay plan can feel good if the question comes up in an exam, so we're ready if our anticipated questions come up.

But they may not – at least not precisely as we imagined. That could spell trouble. We risk being unable to respond or responding in a way that sounds disconnected from the actual question we've been asked. To manage these risks, we need to go back to this diagram. You might remember it from earlier.

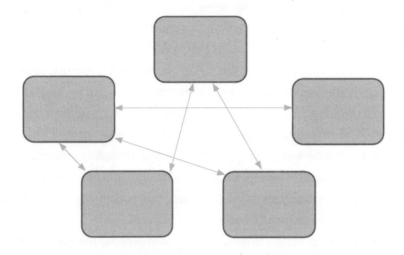

Our goal all along has been to be able to come at subjects via different entry points and using different orders. To this we want to add something else. In the moment, can we construct answers that directly address whatever question we've been asked? That, needless to say, is harder than rolling out pre-planned answers. But because of the work you've already done, it's possible. And part of the reason why, is that there's more time in the questions you're being asked than you might realise.

First of all, let's imagine an interview scenario and a classic question.

'In this new role, you'll be required to manage a large team. Please can you tell us about your experience of managing others in your current role and any previous roles you've held – including specific examples of where you've demonstrated judgement and leadership?'

By my measure, that takes seventeen seconds to say. It's not 'talk about managing others', which takes less than two seconds. But it is essentially the same thing. This is where we can find space.

When I've put in a lot of work anticipating the questions that I might be asked, I become highly attuned to whether they might be arriving. I'm looking for clues from the moment the question starts. In this case, the word 'manage' arrives after three seconds. The question takes seventeen seconds. That is fourteen seconds for me to consider what I've got on the 'shelf' and to decide which order I am going to talk through my strands. Given I've thought about this before, doing that in fourteen seconds will be fine. As they are talking, I'm looking them in the eye and listening closely, but I am also lining up my answer. By the time the seventeen seconds is up, I've had enough to time to be ready to launch into what I want to say.

Picking questions is all about trigger words. When you were fine-tuning your questions list, I recommended settling on one question per subject area. That question can almost certainly be summarised with one or two words. The moment you hear that word, you know, to a large degree, what is going to be required.

If you can start picking questions early, you are giving yourself precious time. You may not think ten seconds is much but if you can stay calm and use it, it gives you a significant head start. Not convinced? Jot down some questions you think you'll be asked on a subject that you've prepared. Select one at random and, after you've seen the question, take ten seconds before answering. At first, you might not make the most of it. Keep going, though, and you'll soon develop the confidence to use the time. It makes a huge difference.

I've learned this from doing the news over the years. When I started, the idea that I'd use ten seconds for anything other than preparing to read the next script would have seemed absurd. As I've become more comfortable in the studio, a ten-second clip is an opportunity for a quick to-and-fro with the editor or to make a change to the next sentence.

Picking the question is the explanation equivalent of picking a tennis serve or a cricket delivery. If you do it, you know ahead of time what's required. There's more we can do, though, to use that time.

SELECTING AND SHARING THE STRUCTURE

As you listen to the question, you'll be thinking about which of your strands to use.

I often think of dynamic explanation like inversions of a chord on the piano. On the piano, you play a C major chord with the notes C, E and G. Or with E as the lowest note, then G and C. Or with G as the lowest note, then C and E. These inversions sound slightly different but they are the same chord. We can think of explanations like this too – there are core messages and information we are going to get across but each time the ordering may be different.

As you listen to a question, these are the seconds where you spot the subject and start deciding what you want to say. Sometimes, you may be so confident in your answer that you launch in without any concerns.

However, if I think there's a risk of me losing my way, I share the structure I've chosen. This has the double benefit of laying out a checklist for me – and of giving whoever's listening a guide to what I'm saying.

Let's continue with our interview example:

'I think I'd highlight three aspects of my career with relation to managing others – my time at X, my previous work at Y and, outside of work, my time volunteering at Z.'

X, Y and Z may well be strands. Once I get into them, I'm going to be confident running through them because they're memorised. Setting out the order in my answer makes it more likely that I get to them. You'll hear effective communicators doing this all the time – it's helpful to them and to their audience.

A variation on this is: 'I think we need to consider four factors when looking at this issue. First . . .' Even without saying what they are – it'll be far easier to remember those four factors and your audience will appreciate that you've indicated a structure and direction. It also demonstrates confidence in how you see the subject, which is always going to increase your chances of people listening.

MIRRORING LANGUAGE

I'm always mindful of appearing disconnected from the question that I'm being asked. So even if I've chosen my answer and am settling on the structure of my response, I'm still listening to the words and phrases that the questioner is using. Here's the interview question again.

'In this new role, you'll be required to manage a large team. Please can you tell us about your experience of managing others in your current role and any previous roles you've held – including specific examples of where you've demonstrated judgement and leadership?'

In this case, 'current role', 'previous roles', 'leadership' and 'judgement' have all been requested. I'm going to use them all.

The strands I've prepared relating to management are now lined up. They will inevitably cover 'current role', 'previous roles', 'leadership' and 'judgement'. Any explanation of my experience in that area would, I hope, have to involve those.

As I line up the strands I've prepared, I'm also thinking about how I can graft the language of the question on to my answers. (This is a technique that is invaluable in exams as well as interviews.)

PULLING IT ALL TOGETHER

Let's imagine how this might go.

> *'I've managed teams in a range of jobs. In my **current role** . . .'* [and I drop on to one strand]

> *'I also showed **judgement** a number of times in my **previous job** . . .'* [and I drop on to a strand about my work there]

> *'I would also emphasise the **leadership** I showed doing . . .'* [and I drop on to a strand about a part of my career that is relevant]

If this goes well, let's consider what we're able to do:

- Create space during the question to start planning the structure of what we say.

- Use strands of information that we've already organised and memorised so all of that work is done.

- Use bridging phrases to allow us to move between those strands smoothly regardless of the order we've chosen for them.

- Mirror the language of the question to ensure our answer directly connects to what we've been asked.

That, I would hope, is going to hit the mark. The example that I've chosen here is an interview. In this, you definitely want to answer the questions directly or you're not going to score so well. However, there are other scenarios where you may really want to make sure you say something – whether you're asked about it or not. Everything we've done so far can help us, as can the next couple of techniques.

SAYING WHAT YOU WANT TO SAY

This takes us right back to that BBC training course of 2002. The message then, as you'll recall, was that there are ways of saying what you want to say regardless of what you've been asked and – I hasten to add – without being rude. The issue being that there will be times when we have information that it is important to say – either for us or the people we're speaking to – and, often through no fault of their own, those asking the questions haven't given us a prompt to do so. You may also be asked a question which you don't have much to say on but want to take the chance to talk about other subjects where you do. This happens all the time to me and I'm sure it does to you too.

Perhaps I'm in a meeting which hasn't got on to one area I think we need to discuss. Perhaps it's a panel at a conference where I really want to emphasise one area of the work we're doing, but no one is asking me about it. Perhaps I'm doing a job interview and I get a question to which I've something to say but not enough for a whole answer.

These scenarios tend to come in two forms – one where you are asked a question that skirts a subject you want to talk about and one where you're asked a question that has nothing to do with what you want to talk about. In both scenarios, we want to shift ourselves into territory that plays to our strengths and prioritises the best information we have to pass on.

Let's take them in turn.

The question that skirts what you want to talk about

If you're asked a question, if at all possible, you need to answer it. The decision then is for how long. Let's imagine we're at a publishing industry conference and we're asked a question about sales but really you want to talk about marketing. In just the same way we outlined above, as the question is coming your way, you pick it. You hear the word 'sales'. In that moment, you reach for your strand that relates to this issue – but you also think this a chance to talk about marketing and you resolve to use two of your strands on that subject as well.

'Yes, sales have been really interesting and encouraging this year . . .' and you start to unpack the strand you have on this.

'I'd also say that we're confident that the performance we've seen in sales connects to the work we've done in marketing. For example . . .' and off you go on one of your marketing strands.

'That's one aspect of our marketing push – another is . . .' and there's your second marketing strand.

'And all of that marketing work has really helped the sales uplift that we've seen in the last year.' And you bring it right back to the question you were asked.

In other words, you start on the question you're asked – you take a detour that is, in part, related to the question – and then you come right back to the question. The calculation here is that while those listening may notice that you've gone on a detour, they won't mind because what you're saying feels relevant and interesting.

The question you don't want to answer

There is a more extreme version of this when someone asks you a question that either you think isn't relevant or you have something else you really want to say before you finish. Or perhaps you actively don't want to answer what you've been asked at all.

If we take those examples in turn.

If you've been asked a question you just don't think is helpful, it'd be rude to point that out. Equally you'll want to move into more productive territory as quickly as you can. To that you might offer a short answer to the question and then say: 'That's one issue. Another I'd mention is . . .'

Perhaps you're at a conference and time is short and you get asked what session you're going to next. No harm in the question but you have more information you'd like to share.

'I think I'm going to the session in Hall A on podcasts. It looks great. And just quickly, before we wrap up, can I mention . . .'

Then there's the question you don't want to answer. I get this quite a lot when being interviewed for articles in the press. Quite reasonably, the journalists are interested in getting my opinions on the state of the BBC or its position on one issue or another. But that's not really for me to talk about as a BBC News presenter (that's one for the bosses . . .). 'I'm not sure I'm the best person to ask on that,' I'll say, before shifting my answer on to territory where I am OK to express myself.

All of these situations are handled using what I call 'escape phrases'. They are a quick form of words that moves you to where you want to be.

If done badly, they can be rude and clumsy. Done well, they can be respectful, helpful and reasonable. Often the people listening to you want to hear what you have to say and won't mind that you've shifted the conversation a little.

Indeed, if you've been asked a duff question, they may be grateful that you have redirected things.

I should add, I know all about duff questions, having asked, I'm sure, thousands of them in my career. I've also learned how to deal with duff questions from the masters of 'escape phrases' – the reporters that I speak to.

Escape phrases and bad questions

Think of 'escape phrases' as performing a similar role to the 'joining phrases' or 'bridging phrases' we looked at earlier. They get you from one place to another. The difference being that with escape phrases you're moving from somewhere you don't want to be, to somewhere that you do.

I have some colleagues who are so good at this, you don't even realise they've done it. They take you from a mediocre question and on to something more important. And even if you spot them doing it, they do so with such fluency and good reason that you don't mind.

My favourite comes from a couple of years ago when I was talking to one of my colleagues in Washington DC. I don't recall the question, but I do recall her answer.

'In some ways, yes. But Ros, another thing is . . .' And within two seconds she was gone! Off my question, which, evidently, was not a good one, and on to much more interesting territory. All of which was completely fair enough. My colleague had limited time to explain a complex story and was ensuring she made the most of that time – for her sake and the audience's. And it wasn't at all rude – in fact, it was done with such assurance, blink and you'd have missed it.

The more I noticed reporters and correspondents using these phrases when talking to me, the more I started noticing them elsewhere. Politicians, business leaders, head teachers, salespeople. Everywhere I looked, the people who were expert at communicating were using escape phrases. I started to note them down and try them out for size. These are some of the ones I've collected.

- *You're right, that is important. As is . . .*
- *I think that's one important issue. Another that ties into this issue is . . .*
- *On that, I'd agree, but if we look elsewhere . . .*
- *It's amazing the number of factors here. That is one – but also think about . . .*
- *You're quite right to raise that. I'd also highlight . . .*
- *One other thing I'd mention . . .*
- *I completely agree. Also . . .*
- *Yes. And that's just one of a number of issues . . .*

And if we're going to be purist about this . . . 'Yes. And also . . .'

These escape phrases are useful to move your explanation to where you want it to be. They also make the prospect of an unwelcome question less daunting. If need be, you can reach for an escape phrase and move to safer waters. And one thing is for certain: unwelcome questions *will* come your way.

ANSWERING QUESTIONS THAT YOU DON'T KNOW THE ANSWER TO

We've just been considering how to answer questions that don't match what we want to talk about. That's not the same as being asked a question you can't answer because you simply don't know what the answer is. That is a *whole* different problem. There are a few different ways I try to deal with this.

Own up

The first is to simply say I don't know the answer. This has the advantage of admitting what is going on rather than the potential awkwardness of hiding it. This is always my preferred option.

You can soften the blow a little with phrases like:

- *Do you know I don't have that – and I'd like to know. I'm going to look into that, but I don't have it to hand.*

- *That's a really interesting question that I'd like to know the answer to too. I'm going to look into that.*

- *That is a detail that I'd like to have to hand. Let me get that after this meeting and I'll send it on to you.*

- *This is something we'd really like to know. It's on our list of things to look into. I just don't have it yet, though, I'm afraid.*

Switch from what you don't know to what you do know

These escape phrases can help you admit you don't know and move away from the issue.

- *I'm afraid at the moment I don't have that. What I can tell you is . . .*

- *So far we don't have those confirmed figures. What is confirmed is that . . .*

- *I don't know the answer to that right now. What I do know is . . .*

All these are honest about what you don't know, but they also get you away from the problem swiftly. These can be even more effective if you can offer something else on the subject.

- *I'm afraid I don't have that particular piece of information. But I do know that on this same issue we've had a number of other updates . . .*

Make a little go far enough

If you have only a couple of pieces of information on the question you're asked, and you really don't want to give an answer that feels underpowered, try to make what you have work for you.

First, you can buy yourself some time. Pad your answer at the front.

> *It's interesting that you ask about this. A lot of people have been raising similar questions – clearly it's got a lot of people talking. Now, on the details of this . . .*

Turns of phrase like this will just give you some breathing room while you consider just how little you know and how you can fashion a decent answer. Then, if you've only two things to say, deliver them, talk around them as best you can to give each as much weight as possible – and then stop.

> *First of all, we know that the policy on this has been updated. It's an update that's got a lot of attention. **It's been decided that X will change**. And we're going to have to wait to see what impact that may have.*
>
> *Second, we're told this policy was changed at **a local level and not a national level**. We're going to be watching closely to see if the national authorities do react.*
>
> *Beyond that, this has only happened today, so a lot of questions remain.*

The highlighted phrases are the only two pieces of information. The rest is a mixture of supporting words that help avoid the answer feeling too thin and an acknowledgement that we'd like to know more.

Lest there's any doubt, this isn't ideal. But if you're in a situation where acknowledging the limit of your know-how is going to be awkward or detrimental, then stretching what you have can help.

In most scenarios, though, I'd prefer a much more straight-forward answer.

> *There are really only two firm updates we have on this. That the policy is definitely going to change and that it was a local decision. Beyond that a lot of people are looking for answers, including me.*

The advice I always give new reporters going on the TV or radio is that if they get a difficult question, answer it as best they can and then stop. A short answer without much information is better than a long answer without much information. And once you stop, another question will follow which may suit you better.

Either way, having *something* is always a lot more comfortable than nothing. And moving off the problem subject, by stopping or shifting, is always better than getting stuck on it.

TAKING CONFIDENCE FROM YOUR PREPARATION

If you've got to this point, you are in a fabulous position to explain yourself. You've had a lot of advice come your way. You won't need all of it every time, but within it there is a range of techniques that brings you to a high level whatever it is that you're doing.

More than that, I'm hoping it gives you confidence to take on the situations you face, however unpredictable they are. I'm hoping too that it helps to ease any nerves you feel. I've long dealt with nerves and, though they certainly get better with experience, they don't go away – not for me, at least.

I still frequently get very nervous, but I've had to face this enough times to know what helps.

By far the most important thing to do is what you'll already have done in the Seven Steps – prepare and practise. I hope you will find great reassurance in the work you've already done. But there are also some other thoughts I keep in mind when I'm offering explanations under pressure.

The first is that people will be hoping and expecting that you are going to be good. There's no reason for them to think otherwise. Whoever you're speaking to won't be on the lookout for disaster. On the contrary, they want to hear what you have to say and would prefer that you say it well. It's the outcome they're hoping for. Assume you're going to be good because they will be assuming that too!

The second is to tune out everything apart from the people you're speaking to. In the end any human interaction is between those involved. The surroundings are certainly worth getting used to if you can – but in the moment, pay them as little attention as possible.

Whenever a contributor comes on to our show and I can tell they're nervous, I say, 'Don't worry about this' – gesturing at the cameras and the lights – 'let's just talk as we normally would.' If they can just see it as a conversation between the two of us, it's going to be much more comfortable for them.

I tell myself the same thing if I'm broadcasting in Downing Street. Right next to me are any number of other presenters, reporters, producers, camera crew and photographers. Behind me people are coming and going through the famous black door, along the street are the police on the gate and, if it's an important political moment, you can really feel the country watching. It would be easy for that to throw you off your stride. To keep myself protected from that, I retreat to what is familiar – talking to the camera a metre in front of me. The same lens I've talked to thousands of times. I try to make it feel like any other day when I talk to a camera. It's the presenter's equivalent of putting blinkers on a horse.

Or to illustrate this another way – the other day, I took part in a session at a journalism conference that was looking at our video explainers. I could feel my nerves coming on. I calmed them by trying to answer each person who asked a question as if I were simply talking to them in the corridor. Again, I was trying to normalise an abnormal situation.

The best communicators are sometimes those who can remain themselves in the strangest of circumstances. Which leads me to my third thought.

When you've done the work, doing what you can do is enough. Sometimes in moments of pressure, we feel a need to be something different, something more, to become a better version of how we normally are. This is definitely not advisable. Time and again, I see people pushing themselves to stretch beyond what is comfortable and beyond who they are, and they rarely communicate or explain themselves better as a consequence. If you've done the preparation, you'll be ready – both for what you hope will happen and what may surprise you.

THE UNEXPECTED IS EXPECTED

Over the years I've done a lot of TV and radio presentation training and we always take time to look at what to do when things start going wrong. Understandably, the new presenters want to avoid an unpleasant experience on air and their managers want reassurance that their new recruits are going to be able to keep the show on the road. And everyone knows that with live TV and radio, it is a case of when, not if, things go wrong.

One of the more stressful thirty minutes of my time on air

was conducting a live TV phone-in from a roof in Johannesburg during the 2010 World Cup – without a functioning earpiece. It had packed up moments before we started and so, as each guest around the world spoke, the quick-thinking cameraman used his hand to signal they were talking and then would signal when they'd stop. I'd then say, 'Thank you', having not heard a word they said, and then, with a 'Now let's bring in . . .', I'd move on to the next guest. I could never stay on a guest for more than one answer because I didn't know what they had said. It was horrible. I was new to TV, already very nervous and, now in this precarious position, I felt exposed and rattled in front of a global TV audience. I don't remember how the show went but I do remember the feeling. The earpiece breaking was a nasty surprise and how I reacted to that had an effect on how I handled the moment.

As I've become more experienced, I've realised that there are many situations we find ourselves in where there's nothing surprising about something not going to plan. This might seem like a small point, but psychologically this can make a lot of difference. If all hell is breaking loose and you're thinking, 'Oh my goodness, what is going on?' you are much less likely to stay cool. If you're thinking, 'Ah, OK this isn't going to plan. Not to worry. Let's see what we can do,' you're going to think a lot more clearly.

I've had some reasonably brutal moments over the years with directors telling me in my ear that I need to fill for minutes on end because all other options have fallen down. Or when technical problems are disrupting our every move. When I was starting out, I could feel myself becoming tense in these moments. These days I don't feel a flicker. Something going wrong is as normal as something going right. That's partly rooted in experience, confidence and preparation. But it's also connected to expecting the unexpected.

This is true of being on live TV, but it's equally true of lots of situations you'll be able to think of. Maybe you're giving a presentation and the slide won't appear. Maybe you arrive for a meeting and there are several attendees you really didn't think would be there. Or maybe, during a conference panel, someone in the audience is unexpectedly hostile towards you. These have all happened to me.

None of these situations would have been the plan. But if you accept that the unexpected could easily happen, when it does, it's not a surprise and you're going to think through the situation much more clearly.

By this stage, you're ready to explain yourself when you control every aspect of what you're saying. You're also ready to do so when you don't. To reach that point is a powerful feeling. It's hard work to get to there but when you do, nerves and doubt are much less likely to bother you. Instead, you'll find a new spring to your step, a conviction to your words and a purpose to how you speak. There is a calmness and authority too. It puts you in the ideal state of mind to communicate what you want to say and explain what you are hoping for from others.

QUICK CHECK

- When you think of the main subjects you need to talk about, are you concerned about any of them?
- Have you gone over your list of expected questions?
- Are you taking confidence from all the preparation you've done?
- One more: do you feel ready? (If you've done all this work, you absolutely are!)

Before we go any further, in case it's helpful to see a quick reminder of the Seven Steps to delivering a great dynamic explanation, here they are again.

SEVEN-STEP DYNAMIC EXPLANATION – QUICK REFERENCE

1. SET-UP
2. FIND THE INFORMATION
3. DISTIL THE INFORMATION
4. ORGANISE THE INFORMATION
5. VERBALISE
6. MEMORISE
7. QUESTIONS

5

QUICK EXPLANATIONS

The Seven-Step Explanation that we've just been through is the explanatory version of throwing the kitchen sink at it. We've already looked at an array of different situations in which you can use it.

There will, though, be many other times when the kitchen sink isn't required or isn't realistic. There is no way that *every* time you have something to explain, you'll be able to work through all Seven Steps in detail. I certainly don't do that.

There are many short written and verbal interactions that are part of the fabric of our day-to-day lives. These aren't moments which we're likely to prepare for – or, if we do, it'll be a snatched two minutes in advance. Here we will rest on good habits that we've learned. But even in those moments it is possible to give ourselves a helping hand. I try to use every opportunity to organise my thoughts and assess the situation whenever I'm having a conversation of consequence. Equally, I'm not going to stop and analyse every email that I send or every short document I write – there simply wouldn't be time. But I am always trying to do certain things when I send short written communications.

QUICK VERBAL EXPLANATIONS

Ahead of any conversation that isn't social, I'd hope to think about what I want to get across, what I want to ask and what I want to learn. This might be for anything from a quick conversation to do with work, to calling the bank with a query, to going to see a neighbour about a local issue.

If this sounds a little over the top, well it is certainly more than I used to do. But it's not the great commitment that it might sound, and it increases the chances of the conversation serving its purpose. That pays itself back many times over in time saved later.

I have three questions, which I quickly try to answer.

1. **Which subjects do I want to discuss?**
    This is unlikely to be a long list, but I want to make sure I cover everything.

2. **What do I want to say?**
    I jot down bullet points under the subjects on my list. Again, this may be a very short list. But there may be a particular point you want to make or a particular piece of information you want to share. You don't want to be kicking yourself afterwards that you forgot to say it. Better to be sure by writing down what you want to say.

3. **What do I want to ask?**
    Write down any questions that you want to make sure you ask.

I normally grab any scrap of paper to do this. You could do it on your phone too. The whole thing can take less than five minutes and often less than one. If a meeting or call happens with even a couple of minutes' notice, I'll try to quickly think through these three questions.

I'm never worried about people seeing that I've prepared what I want to say. As the conversation goes on, I'll keep an eye on my notes. If I've not worked through everything on the list, I'll say, 'Just before we finish . . .'

If for any reason, you don't want to take in your notes, you can quickly do the simplest of the chunking techniques (see page 242) to memorise what you want to say.

These three questions might seem so simple as not to warrant doing. But if you do this in every quick conversation or meeting you have, it can save any number of emails, messages, further conversations and misunderstandings. It makes your conversations more productive and more helpful.

Here it is for reference.

QUICK VERBAL EXPLANATION
– QUICK REFERENCE

1. WHICH SUBJECTS?
2. WHICH INFORMATION?
3. WHICH QUESTIONS?

And if that's quick verbal communication, next there's what we write down.

SHORT WRITTEN EXPLANATIONS

AKA EMAILS & MESSAGES

The rapid escalation in information coming towards us is driven by several dynamics. The proliferation of streaming platforms means there's a lot more TV for us to watch. YouTube and TikTok are driving us to consume vast amounts of video. Podcasts and radio mean there's a lot to listen to. And then there's text. An awful lot of it.

The average person will consume hundreds of pieces of text-based information every day. As we saw earlier (see page 81), email and messages have reached such volume that the social etiquette around them has changed. The most significant development being that there is no expectation that it's possible to read every piece of incoming information or that it's possible to reply. There may have been a time when you sent a message or email and had a reasonable expectation its content would be consumed, acknowledged and responded to. Now you can assume no such thing.

This, I should add, is not a criticism. It's the inevitable consequence of excess information. But if we're to successfully explain ourselves in these day-to-day written environments, we can only do so if we understand and accept the context in which we're operating. It's a context in which we are fighting for attention, in which we will not be given the benefit of the doubt and in which utility matters above all else.

You may think my language has become a little melodramatic. Do emails and text messages really require this much analysis? Do they really matter that much?

Given I'm devoting a section of this book to them, you'll know my answer to those questions. And the reason isn't that I revere email as a literary form or as one of the joys of life. The reason is two-fold.

First, whether we like it or not, this is how a great deal of our communication is done. Institutions, hospitals, companies, schools, sports clubs, airlines and so many other parts of our life communicate with us this way and us with them. In terms of information that impacts our day-to-day life, short written communication is, arguably, the most important. Second, if we don't get it right, the amount of time and effort that can be wasted is extraordinary. Or to be more positive about it, if we explain ourselves effectively from the outset we can reduce the volume of unnecessary communication to a remarkable degree. My wife, Sara, is a tribunal judge and clarity of communication is something she pays close attention to. She calls this 'the initial investment'. I like that. If we pay attention early on, the rewards will follow.

Let's take a hypothetical example. A school is organising a trip for a class. If the email contains all the details the school imagines the parents need and it's well set out, that, for most parents, will be enough. If the email doesn't contain all the relevant information or if it's there but too hard to find, the school will be deluged with questions from parents. There could easily be hundreds of interactions that didn't need to happen. That is a cost to the school, and it may impact how the parents feel about the school. If, rather than a school, this was a business, then there's not just the cost in the wasted time but there may be a cost in lost business or reputational damage.

As such, while individual emails and messages may not feel so important, each one is a chance to not only effectively explain yourself, but also protect your resources.

I'm writing this section of the book while preparing to take part in an event where I don't know the number of people involved, who the people taking part are, who the other guests are, what kind of outcome we're expecting from the day and what kind of space the event is taking place in. I can find all that out by calling the organiser. But I've now had, I think, six emails from them about one aspect of the event or another. So, to avoid looking inattentive, before calling to get answers to my questions, I've had to check over those six emails to see if I missed what I need to know. I hadn't. It's not there. I'm certain, as you're reading this, that you'll be able to think of numerous situations from your life where you've received extensive communication but not received the information you need.

The good news is that making that initial investment in clarity can make all the difference. It can improve your interactions; it can rapidly reduce the time you spend on exchanging emails and messages, and it can enhance your reputation. And we can take inspiration from the guy I saw on Regent Street in London the other day with a t-shirt that read, 'This meeting could have been an email.' That resonated very hard. An effective short written explanation can still cut through the noise and save time. But such is the volume of information coming at us, we need to accept the odds are stacked against us and that needs to shape our approach.

I should add, all my advice here is for communication outside your social and family life. How you message your mum, sister or best mate is entirely between you and them! What I'm interested in is how we interact with people and organisations with whom we have professional or functional relationships.

This isn't only about emails and messages either. Short written communication is a vital feature of many forms of communication from websites and newsletters to pamphlets and posters.

Professor Todd Rogers at Harvard Kennedy School of Government in the US is a behavioural scientist and a specialist in effective communication.

When I say specialist, there cannot be many people in the world who've spent as much time as Professor Rogers studying how we all consume short written communication. (In fact, in the recently published *Writing for Busy Readers*, he and Jessica Lasky-Fink go even further into this.) More important still, Todd has brought his expertise to bear in trying to improve how the state communicates with people – whether to parents about their kids' school attendance or to adults about voting. This, I learned while reading about Professor Rogers, is called 'behavioural policy'. It is the communication strategies that municipal, state or national authorities use. And Professor Rogers' focus is on improving how effective that communication is in order to achieve better outcomes – more people voting, more children going to school, more efficient courts, better understanding all round.

In other words, how does the state explain itself better for the benefit of its citizens? This is explanation with the grandest of purposes.

To answer these vital questions, Professor Rogers has run a variety of experiments, applying behavioural science and data analysis to better understand which types of communication best aid understanding, and which prompt the most engaged responses.

The first time I spoke with Professor Rogers, it was like discovering a kindred spirit. In our own quite different ways – me learning by trial and error through news and my other work, Professor Rogers through academic and systematic study – we are both pursuing how to maximise the possibilities of explanation and communication.

My approach to short written explanations rests on five assumptions – and I've included some of Professor Rogers' advice and research too as it is often complementary.

MY ASSUMPTIONS WHEN I WRITE AN EMAIL

Producing effective short written explanations is hard unless we acknowledge the environment we're operating in and our chances of success. Leaving aside family and friends who (I hope!) are reasonably likely to look at what I send them, here are my five assumptions as I start to write an email.

1. The recipient/s may not read it at all.
2. The recipient/s may not read all of it.
3. The recipient/s will skim it rather than going through it sentence by sentence.
4. The recipient's approach will be entirely functional.
5. If the recipient doesn't feel that it's specifically for them, they are far less likely to read it.

This is true from writing a book or a big speech right down to a quick email. These five assumptions can guide what we produce.

Assumption one: The email or message may not be read at all

This isn't personal. It's not a criticism of the people you're writing to nor a reflection on your importance to them. I'm pretty diligent on messages and emails, but I have days when I simply cannot keep on top of them. On those days, I will still read and respond to some. As such, the emails and messages are in competition with each other. And while I may not always consciously think, 'I'll leave that one and reply to this one', these are the decisions we're all taking as we peruse our inboxes and messaging apps. Based on this, when we send a message or email, we're fighting for attention from the start. Everything we write needs to be geared to convincing the recipient that this is worth opening and then worth reading. We can't be complacent about either.

I do two things to try to help with this. One is to write a subject line that makes clear it is either directly for this person and/or this is about something directly relevant to them.

I also write a first sentence that explicitly explains what the email is about. This is particularly important given some email apps only show the first line or two. It's the same principle I use when writing a news story.

It also always makes me think of 'doorstepping'. This is when a reporter tries to grab a word or two with a politician when they're out and about. Sometimes, such as during an election campaign, there can be a whole throng of journalists and camera crews trying to spot them – and when we do, everyone surges around them. These situations are like a cross between journalism and rugby. You have to battle for a good spot and then, when you're there, you need an opening line that contains just the right information to catch their attention and make them feel inclined to answer. You're competing with many others who are also shouting questions, and you never get more than one short line to make your position count. I think of emails in a similar way!

Ideally, the first sentence needs to explicitly state what the message is about. I'd ditch 'Hope you're well' or 'I hope things haven't been too busy recently' and cut to it.

Hi Joe. I've four questions I'm hoping you can help me with.

Hi Joe. This is an update on next week's trip with all relevant information.

Hi Joe. These are the details you requested on next week's meeting – and I've one question I'm hoping you can answer.

In their 2020 article that I quote right at the start of the book, Jessica Lasky-Fink and Professor Rogers write, 'If a message's purpose is not immediately obvious, readers must allocate more of their limited time to read and understand it. This increases the chance they will just give up and move on to the next item competing for their attention.'

When I spoke with Professor Rogers, I was telling him about my habit of saying what I'm going to do in the first sentence.

'Do you know BLUF?' he asked.

I didn't.

'It's from the US military. It stands for Bottom Line Up Front.'

I laughed in disbelief. I couldn't believe I've been thinking about all this for so long and never heard this before.

Needless to say, I've been using BLUF ever since (and eulogising about Professor Rogers in the same breath). It's a great name for an important tactic.

There is one caveat here, though. This approach is useful within working relationships where being this direct is expected. But not every written interaction will fall into that category. If you don't know someone well, or at all, then it may be appropriate to introduce yourself first and explain why you're writing. We don't want to be so direct that we cause offence. I'll sometimes write, 'I hope you don't mind me messaging out of the blue', 'Great to briefly see you earlier', or 'I'm wondering if I might ask for your help on something'. Although this slightly delays laying out the purpose of your message, it's worth doing if it means getting the tone right.

All that said, with short written communication like email, keep in mind that this isn't a friendly chat in a restaurant or a letter to an old friend. There's no romance here I'm afraid. This is functional and the more helpful and clear you make it from the start, the more likely the reader is to understand the purpose of the email and then keep reading.

Assumption two: They may not read all of it

Again, there's no judgement here. I don't read all of every email that I open. But if I'm sending an email, clearly I'd like people to get to the bottom and there are several things I do to try to address this. You'll notice they are very close to what we've been doing with far more ambitious explanations.

Make the email as short as possible

This may seem obvious, but I receive emails every day that are studies in excess words and sentences. Even reading them feels daunting and, of course, sometimes I don't bother.

Professor Rogers has conducted experiments where he's removed text, even sometimes at random, and the response rate goes up.

In one experiment, cutting the number of words by two thirds increased the response rate by 80 per cent.

It's a simple but powerful equation: make it shorter and people are more likely to read and to respond. If making it shorter is sometimes harder, it's worth the effort for the improved response rate.

Assumption three: The reader will skim

We all skim messages. Often, we're desperately scanning for the information we want. That's all we want, frankly. Rather than fight this, we need to make our emails skimmable. Here's how.

Use short paragraphs

I mean, *really* short paragraphs. Plenty of messages fall at the first hurdle. They are simply too long, contain too many big blocks of text and get closed there and then. They're asking too much of the reader. And before any of us despair that no one can be bothered to read anything of length any more, that is not what is happening here. We can and will read at length. But the vast majority of email is about information exchange, not entertainment or enrichment. As such our commitment to reading email is not emotional or enthusiastic, it's functional. We want the information as quickly and simply as possible.

Short paragraphs really help with this. The space around each piece of information makes it easier to consume.

In a talk in 2021, Professor Rogers explores this. He describes sending a text message to some parents. The first part says, 'Thanks for participating', the second part requests that they fill in a survey. To other parents, he divides it into two separate messages. The request stands alone rather than in the back half of the original message. The response rate when it was split was 15 per cent higher.

Giving important information its own space makes a difference. I would love to know what the response rate would be if the 'Thanks for participating' went second. Pleasantries, while polite, can be obstacles to where the action is. If they can go last, so much the better.

Use formatting

Along with short paragraphs, use headers and bold to tell people what each paragraph contains. Here's an example:

Hi Pri. Updates on three things for you.

1. Details on next week's meeting
It's in room 3 at 4.30 p.m. on Tuesday. I will send the agenda on Monday. No preparation needed from you.

2. New data on last week's product launch
This is arriving tomorrow. I need a day to process it then will send to you and Sam. There is no outstanding data.

3. Advice on approaching Sam
I want to ask Sam about delaying the cuts. Should I raise at next week's meeting or before?

Be careful with formatting. Use it with restraint. There is a severe law of diminishing returns if your email is plastered in bold and different colours. There are other dangers too.

As Professor Rogers told me: 'Don't assume the highlight is benevolent. It decreases the chance of other things being read.' I'd never thought of a highlight as 'malevolent'! But this is of course correct. The point being if too much emphasis is placed on some information, the reader may conclude the rest is non-essential.

You don't always need to write in full sentences

For whatever reason, much of our day-to-day written communication is locked into using sentences. Lots of them. But most of the words in sentences are not what the reader is interested in. They are interested in information. When we're speaking to people, we can't talk in bullet points. That might be a bit extreme! In this kind of written environment, we definitely can. That's an opportunity. Remember, this is a functional exchange of information. For example:

Next week's match
Departure time: 3 p.m.
Leaving from: Bus stop on Jenkins Rd
Match starts: 4 p.m.
Match finishes: Around 5.30 p.m.
Match location: Lloyd's Cricket Club, Lloyd's Rd
Match fee: £5. Pls bring cash
Kit: We have kit for children to use if need be

Or you can even try writing an email like an FAQ page.

When will these changes happen?
We don't know yet. Likely September but the date will be set next week.

How much will they cost?
We've assigned a budget of £500k.

If I want to get involved, who do I ask?
Please contact Sarah.

Will you be announcing this publicly?
Yes. A press release will go out on Monday.

And so on. It can very quickly get people to the information they want.

If you can answer those questions in as skimmable form as possible (I'm not sure if 'skimmable' is a word but I'm going to go for it), then your chances of explaining what you're doing will be much higher.

Assumption four: This is a functional exchange

Coming back to my man with the 'This meeting could have been an email' t-shirt, this is an information exchange. It's not only about you passing on information, but also, in many cases, you wanting information from someone. And often the person I want information from is very busy. Many managers I deal with (at the BBC and elsewhere) are in constant motion – moving from one meeting to another, often with little space between them. I need information or a decision from them, but it doesn't warrant setting up a meeting and they're not going to have time to read a long email. I'll acknowledge this and explain exactly what I need. I'll make it as easy as possible for them to help me. Or in the words of Professor Rogers, 'provide all information needed to act in one place.' For example:

Hi Susan. I've three quick questions. Once I know the answers, I can move the project to the next stage, as we discussed.

1. The last quarter revenue was as projected. Are you happy for me to set the budget without you seeing it again?

2. Do you want to come to the meeting with Len on Thursday at 2 p.m.? It is only discussing the budget.

3. Can we talk in public about this yet? Am getting mixed messages from press, marketing and sales.

Often when I send an email like this, I'll get a reply where the recipient has put very short (sometimes one word) replies under each. The clarity allows me to get on with my work, and I've used up the minimum of their time. Always stay focused on what you want to explain to them and what you want them to explain to you.

Assumption five: Group messages and emails reduce your chances of communicating and of receiving a reply

I talked about this earlier in the book. The more you can create a sense that 'This is useful to *you*', the more likely the recipient will engage.

I am going to use how we distribute our explainer videos as an example of how I try to do this. The first thing I do is to avoid using group emails if at all possible. Our explainers are taken by a range of BBC outlets though no one is obliged to take them, so every time I've a new one, I need to make the case. I will email each of those outlets separately. I can tailor my offer to them according to their needs. They get the most relevant information to them. It also demonstrates that I'm committed to making it work for them. Sending ten emails will take more time than sending one, but engagement will be so much higher that it is time well spent.

Once I've heard back from everyone, I've a list of who wants the next explainer. We may then email them as a group – but we continue to make it as easy as possible to find the information that is relevant to them.

Here's an example of how this can work if you're emailing a range of people – say, about a seminar that they're all involved in.

Hi everyone. Here are full details of the seminar, including instructions specific to your roles.

Location: The auditorium
Time: 1–2 p.m.
Audience arrival time: 12.30 p.m.
Audience size: 150
Make-up of audience: Mixture of undergrad students and colleagues

If you're a speaker . . .
Arrival time: 12 noon
Your presentation: Pls send by 12 noon the day before
Point of contact: Simon – cc'ed on this email

If you're in the technical team . . .
Arrival time: 11 a.m.
Your passes: Pls bring them or you can't get in
Point of contact: Jane – cc'ed on this email

And so on. This means that even though you are receiving a group email, the recipients can still feel it is for them.

What I've done here is not dissimilar to a 'call sheet' that you get in lots of theatre, TV or conference productions. However, I'd advocate using this approach well beyond the organisation of events or productions.

The more you distil and lay out the information, the more you answer people's questions, the more you make your explanation a utility, the more likely people are to read it and understand it.

Just as importantly, the more you reduce unnecessary communication with people whose questions you've not answered.

It's possible there are more exciting forms of explanation, but there aren't many that can have such an impact on our lives. For better or worse, this is how much of our day-to-day communication comes at the moment. We all need to explain ourselves well in this form because often there isn't an easy alternative. If we can do that, it can save time, it can increase the chance of a response, it can increase the chance of you being understood and it can make the recipient feel positively about you (you're saving them time by sending well-explained messages).

'NO ONE CARES ABOUT MY EMAIL AS MUCH I DO'

The afternoon that Professor Rogers and I talked, I was almost vibrating with interest in what I was hearing. His work is so relevant to much of what I've been thinking about for twenty years. And he left with me with two phrases that are invaluable whenever we undertake any short written communication – email especially. The first is the one in the header: 'No one cares about my email as much I do.'

Has a truer word ever been uttered? This is on point and, really, this can apply to any explanation that you're offering.

Just as when I make a video I assume no one is going to watch, with short written communication, a decent working assumption is that it won't be read. Once you've made that assumption, you can work to make sure that's not the case.

Then there's the phrase Professor Rogers uses to describe a long email. He calls it 'an unkind tax.'

I love this too. I'd always thought of long emails as exercises in ineffective communication – where the author loses out by not communicating effectively. I'd never thought of long emails actively creating a burden on the recipient and taking time away from them. Whenever I get a long email now, I think of Professor Rogers' 'unkind tax'.

It's not just for selfish reasons that short emails are good – they are good for everyone just as all clearly explained, short written communication serves both the writer and the reader. Their immediate task is information exchange but beyond that lies an infinite amount of actions and outcomes that may not occur without that initial communication doing its job. As I've gone through my career, I've felt that more and more. My interest in emails directly correlated with the difference I could see them making. Each day we open and close possibilities according to how well we write to people.

SHORT WRITTEN EXPLANATION
– QUICK REFERENCE

1. EXPLAIN THE MESSAGE IN THE FIRST SENTENCE
2. IS YOUR MESSAGE AS SHORT AS POSSIBLE?
3. IS IT FORMATTED TO BE SKIMMABLE?
4. IS IT EASY TO RESPOND TO?
5. HAVE YOU ANSWERED THE READER'S QUESTIONS?

CONCLUSION

THE ART OF EXPLANATION

A few months back, Professor Jeff Jarvis of the Craig Newmark Graduate School of Journalism in New York was kind enough to tweet: 'The BBC's Ros Atkins has made a practiced art of explanation.' Aside from, of course, being flattered, one word caught my eye. *Practiced* (or 'practised' if you prefer the British spelling . . .). Because, as I hope has come through in this book, explanation is a multifaceted endeavour. It evolves and adapts according to circumstance and according to how attuned you are to what's required. For me, some aspects of good explanation will always require processes to support them. Much of what I've described here, I do every day. Over time I've got quicker and more confident using these techniques and some are well-grooved habits now. But I don't run on instinct alone. The rigour of working through the steps and of working through the checklists are, for me, still necessary. They help get all of the dimensions of a good explanation as in sync as possible. There is an art of explanation – but it's one you have to work at. You have to practise.

How you practise, though, is down to you. Just as with a cookbook, you wouldn't try to do all the recipes within a week of buying it. Similarly, this, I hope, is a book that offers you a range of starting points.

Perhaps pick one aspect of how you communicate in your day-to-day life and see if you could start there. It might be how you email, how you approach meetings with colleagues or how you interact with clients. Or perhaps there is one important moment on the horizon – a presentation, an interview or an essay – and you could use that as a test to try the full Seven Steps. And, just as whenever we learn something new, each time it'll become more familiar and, fingers crossed, each time it'll be more effective too.

There is a thrill to doing explanation well. In part, from the sense of achievement in taming complexities, of finding a way to handle and shape information in unpredictable environments. In part, for the sheer joy of getting to the essence of something, of knowing *this* is what you want and mean to say. And *this* is how you want to say it. But if there's a thrill – there's a shock too in understanding how much of what we say can undermine what we're hoping to do. Once you see and feel that, it's very hard to forget it.

All our lives are shaped, in part, by the information we share, the information we seek and the information we receive. They're shaped too by our interactions with family, friends, colleagues, officials and many others. How we explain ourselves can impact all of that. It can improve our lives and the lives of others. It can create time and space that goes to waste when information is not clearly explained. It can help us understand and communicate on subjects which feel overwhelming. It can help us speak with confidence and clarity in fluid and intimidating circumstances. It can bring success and change. And even when it doesn't, we know at least we've done what we can. To use a familiar phrase one last time: good explanation gives us the best chance. For me, that is always enough.

While writing this book, I rang my colleague Allan Little. I told him how much I'd taken from his training course in the BBC archive and we exchanged notes on why we're both pre-occupied with all of this. Why spend this much time fretting over precisely what we say and how we say it?

'It's like cleaning a window,' Allan reflected. 'You can see through a dirty window but if you clean the window you can see so much better.'

'I'm going to use that,' I joked with him. And so I have. Because that is what this book is about and it's why explanation matters. We can all get by without paying attention to how we communicate, but when we do pay attention, the possibilities open up in front of us. 'Cleaning the window' means we see out and the world sees in. Once you start doing it, there's no going back.

References

1   www.bostonglobe.com/2020/12/19/opinion/write-shorter-messages/

2   www.journalism.co.uk/news/new-york-university-and-propublica-launch-explanatory-journalism-initiative/s2/a541810/

3   Kueng, Lucy, *Hearts and Minds: Harnessing Leadership, Culture, and Talent to Really Go Digital* (Reuters Institute for the Study of Journalism/University of Oxford, 2020)

4   https://www.theguardian.com/world/2022/sep/22/new-zealand-hopes-to-banish-jargon-with-plain-language-law

5   Tweedie, Steven, 'Steve Jobs Dropped the First iPod Prototype into an Aquarium to Prove a Point', *Business Insider*, 18 November 2014

6   Brown, David W., 'In Praise of Bad Steve', *The Atlantic*, 6 October 2011

7   www.bbc.co.uk/bitesize/guides/zbmrd2p/revision/5

8   www.forbes.com/sites/pragyaagarwaleurope/2018/08/15/why-brand-stories-matter-and-simple-steps-to-create-a-unique-brand-story/

9   www.thoughtco.com/common-redundancies-in-english-1692776

10 Syed, Matthew, *Bounce: The Myth of Talent and the Power of Practice* (HarperCollins, 2010)

Acknowledgements

As will become apparent when you read this book, I'm keen on asking for advice. As such, I'm acutely aware that however many people I thank and reference in the book, there will be many others to whom I owe a great deal. There are, though, some people without whom I wouldn't have a book in which to write these acknowledgements and I would like to mention them here.

At school, my English teacher, John Scott, and history teacher, Jane Rainbow (or Miss Thomas as we knew her then), both infused me with an enthusiasm for thinking about how we consume, organise and communicate information. The same was true at university, as my studies were guided by the inspirational historian Professor Keith Wrightson.

Much more recently, at the BBC, I'm indebted to a number of editors including Liz Corbin, Liz Gibbons, Jess Brammar, Dominic Ball and Finlo Nelson Rohrer. Just as I am to producers Michael Cox, Mary Fuller, Andrew Bryson, Harriet Ridley, Briony Sowden, Courtney Bembridge, Ben Tobias, Eliza Craston, Catherine Karelli, Ellyn Duncan, Tom Brada, Jack Kilbride and Rebecca Bailey, with whom I've worked and worked on how we can better explain the news. I'd be lost without them. Just as I would three of my dearest colleagues: Simon Peeks, Fiona Crack and Nuala McGovern. Whenever I'm hatching a plan, they tend to get a call and I'm always the better for doing so. The same is true of my friend Vivian Schiller, and my agent Miranda Chadwick.

When it comes to this book itself, thanks to Dr Aneeta Rattan of London Business School and Siri Chilazi of Harvard Kennedy School for suggesting I could write something before it had crossed my mind. Thanks to my friend Victoria Moore who was the person I called when I decided to give it a go and to Laura Barber at Granta for her advice too. Each step of the process, people have generously helped me to feel my way.

So it has been with the brilliant team at Wildfire, in particular Publishing Director Alex Clarke and my ever-patient editor Lindsay Davies. I've felt in the safest of hands throughout. Thanks too to my literary agent Will Francis at Janklow & Nesbit who's expertly guided me through this new experience from start to finish.

I must also mention my friend and former BBC colleague Jonathan Yerushalmy who critiqued one of the early drafts. My mum and sister Ismay would take on later versions and they, along with my dad and sister Jen, have been constant in their support and interest. They've helped both to make the book better and, just as importantly, to get it done.

Finally, thanks to the three people most affected by this book. Getting something like this together while still working full-time inevitably impacts life at home. My two daughters, Alice and Esther, will have lost track of the times I've said, 'I'd love to, but I've got to get on with the book'. I'm already enjoying being able to say 'I'd love to' with fewer caveats.

Then there's my wife, Sara, who has both taken the weight when I couldn't and who's offered advice, encouragement and perspective when I've needed it – along with, I should say, some very good ideas on explanation too. Put simply, without Sara's help, I couldn't have done this. I hope, if I've managed to explain anything while writing this book, it's how grateful I am to her for that.

Ros Atkins
September 2023

Index